MINNESOTA
In Bloom

MINNESOTA In Bloom

Find and Identify the 142 Most Common Wildflowers

Michael Homoya

TIMBER PRESS
PORTLAND, OREGON

Frontispiece: Showy lady's slipper (*Cypripedium reginae*)
Opposite: Prairie onion (Allium stellatum)

Timber Press
Workman Publishing
Hachette Book Group, Inc.
1290 Avenue of the Americas
New York, New York 10104

timberpress.com

Timber Press is an imprint of Workman Publishing, a division of Hachette Book Group, Inc. The Timber Press name and logo are registered trademarks of Hachette Book Group, Inc.

Printed in Dongguan, China, (TLF) on responsibly sourced paper

Text layout by Mary Velgos, based on series design by Hillary Caudle
Cover design by Leigh Kaisen

Endpaper illustration by Alan Bryan

ISBN 978-1-64326-460-8

A catalog record for this book is available from the Library of Congress.

To my brother Chris and sister Rebecca

Contents

9 Preface
11 Acknowledgments
12 Getting to Know Minnesota
20 Plant Species and Names
36 Conservation of Native Plants and Their Habitats
38 Using This Book
43 ● White Flowers
147 ● Pink to Red Flowers
187 ● Orange Flowers
197 ● Yellow Flowers
259 ● Green Flowers
271 ● Blue to Violet Flowers
329 ● Brown to Maroon Flowers
340 Interpreting Scientific Names
343 Bibliography
344 Photo and Illustration Credits
345 Web Resources
345 Native Plant Societies
346 Index

Almost every person, from childhood on, has been touched by the untamed beauty of wildflowers.
—Lady Bird Johnson

PREFACE

Minnesota, "Land of 10,000 Lakes," has much more than picturesque lakes. If you look closely, you'll see that it is also home to a wide array of interesting and beautiful wildflowers. More than 1750 native species of vascular plants have been recorded here, plus about 450 additional ones that have been introduced and naturalized from elsewhere in the world. There are species of every size, shape, and color, along with nearly 50 species of native orchids, including the intricately flowered calypso and the strikingly beautiful showy lady's slipper, the state flower of Minnesota.

Wildflowers are most common in natural areas, but even if you have never set foot in such a place, you may already know, or at least be familiar with, some of these plants, such as wild species of iris, geranium, and lily. Of course, other wildflowers will not be so familiar, such as the very rare kittentails (*Besseya bullii*) and dwarf trout lily (*Erythronium propullans*). There are bizarre plants too, such as dodder (*Cuscuta* species), a leafless parasite that looks like yellow-orange strands of cooked spaghetti, and the carnivorous purple pitcher plant (*Sarracenia purpurea*), whose vaselike leaves contain deadly pools that trap and digest helpless insects. Whether they are colorful, showy plants or the bizarre and deadly, these and more are presented here in this easy-to-use field guide that features text and photographs of some of Minnesota's more common and interesting wildflowers. (For a treatment covering many additional wildflower species found in the state, consider *Wildflowers of the Midwest* by Homoya and Namestnik.)

Why learn about wildflowers? One incentive is that they are good indicators of natural quality, and the higher the quality, the better the health of the land (and you too!). But if for no other reason, learning to identify plants (and animals) helps you "avoid or overcome the monotony of anonymous scenery," as my botany professor Dr. Robert Mohlenbrock wrote so aptly years ago. It is my hope that this book will not only help make Minnesota's natural scenery less anonymous to you, but it will also entice you to cherish and protect it. As essayist Aldo Leopold wrote, "When we see land [and the associated wildflowers] as a community to which we belong, we may begin to use it with love and respect."

ACKNOWLEDGMENTS

No book is ever the effort of only one person, and that is certainly the case with this one. I am greatly indebted to the many people who helped me in so many ways, from tracking down information about a species' occurrence to reviewing the draft. For all their help, I wish to thank Derek Anderson, Michele Angel, Harvey Ballard, Donna Baron, David Boufford, Francis Bremer, Teresa Clark, Moira Fitzgerald, Ryan Harrington, Daniel Hinchen, Jacoba Lawson, Adrienne Mayor, Will McKay, Sarah Milhollin, Julia Morrow, Jessica Murphy, Scott Namestnik, James Pringle, Andrea Rapacz, Tony Reznicek, Pat Schaefer, Jeannie Sherman, Welby Smith, Reiner Smolinski, Lisa Theobald, and Gerould Wilhelm. It is much appreciated.

Of course, this book wouldn't be what it is without the photographs. I took some of them, but most of the photos were captured by others, and for the use of them I offer my sincerest "thank you" to Adam Balzer, Katy Chayka, Peter Dzuik, Andrew Gibson, Peter Grube, Karen Hoksbergen (for use of the image taken by Dan Tenaglia), Michael Huft, Eric Hunt, Scott Namestnik, Nathanael Pilla, Corey Raimond, Paul Rothrock, Perry Scott, Brad Slaughter, and Steve Turner.

My lovely wife, Barbara, was a tremendous help and inspiration. She contributed in so many ways, including reading and editing several of the drafts and, perhaps most importantly, providing moral support (and patience) during the research and writing phase. Our two sons, Aaron and Wesley Homoya, were invaluable whenever I needed help with computer issues. Thank you all, dear family.

Getting to Know
Minnesota

I discovered some of Minnesota's beautiful landscape during my first trip to Superior National Forest and the Boundary Waters Canoe Area Wilderness. I observed, among other things, pristine lakes and wetlands and fantastically rugged topography, including Eagle Mountain, the highest point in the state. Add in expansive forests, peatlands, and rolling prairie within Minnesota's some 87,000 square miles, a small portion of which I have had the pleasure to botanize, and it's no wonder to me that the state has a wealth of wild lands and wildflowers. Describing all the natural features that Minnesota offers would need its own book. I have simplified the discussion by dividing the state into two ecoregions: the Northern Lakes Ecoregion and the Tallgrass Prairie Ecoregion.

The Northern Lakes Ecoregion, which covers most of the northeastern half of the state, owes much of its appearance to glaciation. During the Pleistocene Epoch several millennia ago, the region was trounced by several advances of massive ice sheets that measured up to a mile thick in places. The resulting landscape we are graced with today comprises a complex of forests, picture-perfect lakes, moraines, till plains, and lake plains. This ecoregion is also where we find the greatest occurrence of peatlands in the state, most in the form of marshes, swamps, bogs, and fens. These saturated peatlands are commonly quite open, but many are forested, with distinctive trees such as black ash, paper birch, yellow birch, white cedar, balsam fir, red maple, black spruce, and tamarack. On the better-drained sites, we find bigtooth aspen, quaking aspen, American basswood, black cherry,

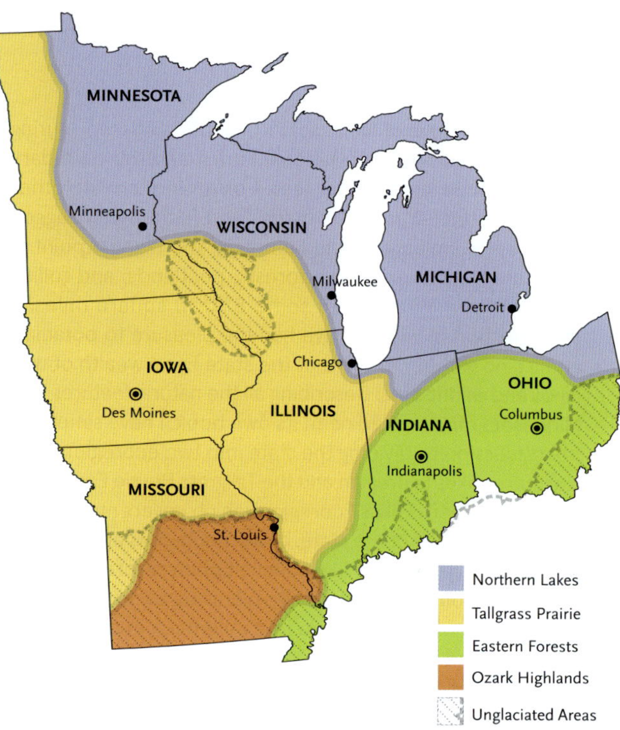

sugar maple, northern red oak, and white pine. Jack pine and red pine are characteristic of drier sites.

Most of the southern and western parts of the state are included within the Tallgrass Prairie Ecoregion. Prairie, savanna, barrens, and woodlands as well as some areas of deciduous forest characterize much of this area, or at least once did. A significant portion of those natural communities has been altered or eliminated by land conversion or fire suppression to the degree that only remnants remain. The forests are scattered but occur especially in areas along watercourses or hillsides too steep to cultivate. Some of the typical trees include green ash, quaking aspen, American basswood, box elder, American elm, hackberry, bur oak, American plum, and peachleaf willow. Many plant species, including some of these trees, are on the western edge of their natural range here, and as one travels farther north and west, their presence drops somewhat precipitously.

There is a transition zone where the two ecoregions meet. When looking at forest composition, the zone represents a line where coniferous forests meet those dominated by broadleaf deciduous trees. Note that there is considerable overlap—a "mash-up" if you will—of species and natural communities between the two ecoregions. Nature does not usually abide by the lines and boundaries we assign to it.

A special landform occurs in the far southeastern corner of the state. Known as the Driftless Area, it consists of land mostly untouched or little-touched by continental glaciation and consequently is composed of rugged hills, valleys, and deep-cut, rocky gorges and cliffs. It is home to a variety of plants, some of which are found in the state only here.

Where to look for wildflowers

Fortunately, there is no shortage of places in Minnesota to look for wildflowers. National and state forests, lakeshores, nature preserves, parks, scenic riverways, and wildlife refuges are all excellent places to explore. All combined, many thousands of acres of public lands are in the state. Private property offers even more areas; just be certain to acquire permission before visiting.

Public properties are organized here by the ecoregion in which they occur. Readers are directed to contact the owners or site managers for specific up-do-date information before visiting properties of interest. Some of the sites may be remote, especially the scientific and natural areas, and do not have maintained trails or other recreational facilities.

Northern Lakes Ecoregion

- Chippewa National Forest
- Foot Hills State Forest
- Itasca State Park
- St. Croix National Scenic Riverway (shared with Wisconsin)
- St. Croix State Park
- Superior National Forest
- Tettegouche State Park
- Twin Lakes Scientific and Natural Area
- Voyageurs National Park

Tallgrass Prairie Ecoregion

- Blue Mounds State Park
- Buffalo River State Park
- Bluestem Prairie Scientific and Natural Area
- Northern Tallgrass Prairie National Wildlife Refuge
- Oronoco Prairie Scientific and Natural Area
- Pembina Trail Scientific and Natural Area
- Prairie Coteau Scientific and Natural Area

Natural Communities

Wildflowers are where you find them. But finding them doesn't need to be a random exercise. For the most part, plants occur in specific natural community types (habitats). Such is where organisms live, interact, and obtain resources to survive. Thus, if you're looking for a particular wildflower, you'll find it advantageous to seek out and locate the natural community in which it is known to grow. Natural communities are often classified by dominant vegetation, soil type, moisture condition, and topographic position. The state of Minnesota includes several general natural community types, which are referenced in this book as barrens, bogs, fens, forests, lakes, marshes, prairies, savannas, springs, streams, rivers, swamps, and woodlands, as well as disturbed sites created by humans.

BARRENS A usually dry, nutrient-poor site with sparse vegetation. If trees are present, they are often stunted and twisted. Sites are often excessively drained.

BOG A typically open and acidic, nutrient-poor wetland, commonly with a floating mat of living and decaying sphagnum moss dominated by heath family shrubs. Usually in a basin, not fed by groundwater.

DISTURBED SITE An area that has been disturbed mostly by human activity, such as agricultural fields, industrial sites, rail yards, ditches, roadsides, and wastelands. Plants growing in these sites are often invasive exotics, but some native species such as common milkweed (*Asclepias syriaca*) may thrive here.

FEN A site with highly organic soil saturated with groundwater, usually flowing in a slow, diffuse manner. Fens can be alkaline, circumneutral (nearly neutral) or acidic, open and graminoid (grasslike), or forested.

FOREST A site dominated by trees that usually have tall, straight trunks and commonly form a closed upper canopy. Layers of smaller trees and shrubs form midstory and understory. Occurs on sites ranging from very wet (swamp) to xeric (extremely dry).

LAKE A relatively large body of water that is often sufficiently deep that only aquatic plants can survive. Ponds have similar characteristics but are smaller in size.

MARSH A wetland dominated by sedges, grasses, cattails, and rushes. The water level is usually relatively shallow.

PRAIRIE A mostly treeless area dominated by native grasses and forbs (nonwoody, non-grasslike flowering plants).

SAVANNA An area dominated by grasses and forbs but interspersed with widely spaced trees. Generally, less than 30 percent of the canopy cover is trees.

SPRING A natural, continuous or intermittent flow of water that emerges from the ground.

STREAM/RIVER Flowing waters confined within channels and banks occurring in valleys and floodplains. Associated communities include gravel bars, sandbars, mudflats, and banks.

SWAMP A forested or shrubby wetland, usually with standing water, the source of which is direct precipitation, flooding, or groundwater seepage.

WOODLAND Similar to forest, but trees are typically less dense and do not form a closed canopy. Midstory and understory are usually sparse because of substrate conditions and/or fire.

Plant Species and Names

People have been naming organisms since time began, but it was 17th-century English naturalist John Ray who first defined "species" as the basic level of classification and naming. Generally speaking, a species is a group of organisms that share physical and genetic characteristics and can interbreed and produce fertile offspring. Species can have two kinds of names: a common name and a scientific name. Common names can be interesting but are also messy, because many plants have more than one common name, and a problem arises when a plant is known by one name in one area and a different name elsewhere. For that reason, I highly recommend that you learn scientific names, which are Latin or Latinized words. They are standardized, and regardless of where in the world you live, the scientific name for a particular species is the same. It consists of two parts: the genus and specific epithet. The combination of the words is known as a binomial. Don't be intimidated by the Latin. You may not realize it, but you are likely already familiar with this naming convention. For example, you probably know *Tyrannosaurus rex* and our own scientific name, *Homo sapiens*. Now consider one of our more common wildflowers in the state, the northern blue flag. *Iris* is the genus name and *versicolor* the specific epithet. Together they form the binomial scientific name *Iris versicolor*. Another native iris in Minnesota is the southern blue flag (*Iris virginica*). Because these two plants share the same genus name, we know that they are obviously related, but the different specific epithets tell us that the two are different species and to a degree are consistently different from each other.

Plant Family Classifications

A plant family represents another level of classification that groups related plants together. Families are based on similarity of reproductive structures, typically the flower and fruit. Today, due to advances in technology, family designation is also determined by an organism's genetic makeup. Just like human families, plant families show relationships. Often it is apparent that a particular plant, even if it's initially unknown, belongs to a particular family. For example, one trait of members of the carrot family is that their flowers are commonly arranged in an umbel (in which flower stalks spread from a common point, kind of like an inverted umbrella). Although some plants with umbels belong to other families, identifying an unknown plant having an umbel should at least suggest that it *could* be a member of the carrot family.

Plant family characteristics

The characteristics provided in these descriptions are focused on species belonging to the plant families treated here. A particular family's overall diversity, however, may display other characteristics or growth forms that are not mentioned for the Minnesota species. For example, some violet species grow as shrubs (in Hawaii!) but do not do so in Minnesota. In some of the accounts, houseplants or plants of the garden or grocery produce section are mentioned to show relationships between our wildflowers with plants that may be more familiar to you.

Sweetflag family (Acoraceae)
This small family of herbaceous wetland plants features long, sword-shaped leaves with parallel veins, somewhat like those of iris, that produce a sweet odor when bruised.

Flowers are inconspicuous and regular, with six tepals (sepals and petals indistinguishable), numerous on a cylindrical club (spadix). The species were formerly placed in the arum family (Araceae).

Water plantain family (Alismataceae)
These herbaceous perennial plants grow in water or mud. Leaves are usually simple and entire. Flowers have three white petals, arranged on a stalk with a whorled branching pattern on a terminal flowering stalk.

Onion family (Alliaceae)
These perennial herbaceous plants usually grow from bulbs, with simple narrow leaves (or broad in wild leek), mostly basal, sometimes alternate. Most have the characteristic onion or garlic scent when bruised. Flowers are regular with six tepals, usually in a rounded cluster atop the flowering stalk, sometimes replaced by bulblets.

Carrot family (Apiaceae)
Species in the carrot family are usually easily recognized by their flat-topped or rounded umbel of flowers. Leaves are often highly divided and have sheathing bases that wrap around the stem. Flowers are small with five petals. In addition to our wild plants are several important garden plants in the family, including carrot, parsley, cilantro, parsnip, dill, and others.

Dogbane family (Apocynaceae)
This family includes milkweeds (*Asclepias* species, which were formerly in their own family, the Asclepiadaceae), and they outnumber dogbane (*Apocynum* species) in Minnesota. Plants are perennial and herbaceous with simple, entire leaves and milky sap in most species. Flowers have five petals (those of milkweeds also have five tubular hoods, each with a horn), in

clusters atop the plant or in leaf axils. Seedpods are swollen and lance- or narrowly egg-shaped. Seeds often have silky hairs at the tips to help disperse them in wind.

Arum family (Araceae)
These herbaceous plants are characterized by having a spathe (hood) and a spadix (club) upon which inconspicuous flowers are attached. Flowers lack petals. The plants contain calcium oxalate crystals and thus are not to be eaten. This family is famous for its many tropical species, many used for their interesting foliage as houseplants, such as heartleaf philodendron, red heart anthurium, dumb cane, and pothos.

Ginseng family (Araliaceae)
This family of perennial herbaceous plants and shrubs have mostly alternate, simple or compound leaves. Flowers are small and regular, with five petals in an umbel or cluster of small umbels, often producing a somewhat flattish array of flowers. The highly valued ginseng is harvested from the wild as well as being cultivated.

Pipevine family (Aristolochiaceae)
The leaves of these perennial herbaceous plants are mostly heart- or kidney-shaped and entire. Flowers with three petal-like sepals form an unusual shape, somewhat like a miniature pipe or urn. Elsewhere, several members of the family are vines, such as Dutchman's pipevine that mostly occurs farther south.

Aster family (Asteraceae)
The Asteraceae is considered one of the largest plant families in the world and includes thousands of species. It was originally known as the composite family (Compositae). Flowers are packed together on a specialized head that often results in an impression of a large, single flower (think sunflower).

Two types of flowers are commonly present on the head, disk flowers and ray flowers, though in many species there is just one type. Disk flowers are typically tubular in shape, tiny, and packed together on the disk in the center of the head. Ray flowers are much larger and often flat, and they encircle or completely cover the disk. The base of the head is covered with tiny leaflike bracts.

Touch-me-not family (Balsaminaceae)
Our species of *Impatiens*, the local representatives of the touch-me-not family, are herbaceous annuals with juicy, fleshy stems and simple leaves. Flowers are irregular, usually with three petals and three sepals, the lower one with a nectar spur. The fruit is typically a capsule that, when ripe, explodes upon touching or other disturbance to expel its seeds. Exotic species are popular bedding plants, such as New Guinea impatiens.

Barberry family (Berberidaceae)
Native members of this family in Minnesota are perennial herbs with simple or compound leaves. Flowers are regular, usually with six petals or petal-like sepals. The fruit is a berry (in mayapple) or a berrylike seed (in blue cohosh). Japanese barberry (*Berberis thunbergii*), a shrub, is commonly used in landscaping and is also quite invasive.

Borage family (Boraginaceae)
These plants, often bristly-hairy, have simple basal or alternate leaves with regular, somewhat tubular flowers with five lobes. They are commonly in coiled clusters that unfurl as blooming commences. The true forget-me-not (*Myosotis scorpioides*) is nonnative and quite invasive in several moist habitats.

Mustard family (Brassicaceae)
This large family consists of several annual, biennial, and perennial species that often have pungent, watery sap. Leaves are alternate and/or basal, simple or deeply pinnately lobed. Flowers are typically regular, with four mostly white or yellow petals. Fruit is a distinctive, mostly narrow capsule that, when mature, splits into two halves to expose the seeds. Also known as the Cruciferae—a reference to the crosslike arrangement of the petals—still a legitimate name for current use.

Watershield family (Cabombaceae)
These perennial aquatic plants have alternate floating leaves. The leaf stalks are attached to the central portion of the blade. Flowers typically are regular with three petals, usually occurring singly on long stalks at or just above the water's surface.

Bellflower family (Campanulaceae)
Members of this family of annual and perennial herbaceous plants usually have simple alternate leaves and often contain milky sap. Flowers are regular or irregular, usually united at the base and then tipped with five lobes. Although lobelias are considered members, some botanists place *Lobelia* species in their own family, the Lobeliaceae.

Pink family (Caryophyllaceae)
The leaves of this family of annual and perennial herbaceous plants are usually opposite, simple, and entire. Flowers are regular, usually with five petals (rarely four or petals absent) that may be lobed, often with five or ten stamens, one to many in branched flower clusters. Also known as the carnation family, it includes chickweed as well as carnation and baby's breath of the florist trade.

Spiderwort family (Commelinaceae)
These herbaceous annual and perennial plants have leaves that are simple, entire, and alternate, with a sheathing or clasping stem. Flowers are regular or irregular, with three petals, in a branched or umbel-like cluster, often with a bract just below. Most species are tropical.

Lily-of-the-valley family (Convallariaceae)
These perennial herbaceous plants were formerly included in the lily family (Liliaceae). Leaves are basal, alternate or whorled, simple, and entire. Flowers are regular, with four, six, or eight tepals, solitary and axillary or sometimes branched in terminal or axillary clusters. The frequently cultivated lily-of-the-valley (*Convallaria majalis*) is a European native plant.

Morning glory family (Convolvulaceae)
This family includes annual and perennial herbaceous plants that are typically vining or trailing. Leaves are usually simple and alternate, often lobed. Flowers are regular and usually bell-shaped, with five mostly shallow lobes, solitary in leaf axils or in clusters.

Gourd family (Cucurbitaceae)
These annual and perennial herbaceous vines usually have coiled tendrils. Leaves are usually alternate and palmately lobed. Flowers of separate sexes are regular and usually five-lobed. Fruit is usually a berry or pepo (fleshy with a hard, inseparable rind). Common garden plants in the family include cucumber, watermelon, and squash.

Sundew family (Droseraceae)
This is a family of very specialized carnivorous, perennial, herbaceous plants that usually consist of a basal circle of

modified leaves whose surfaces are covered with sticky, gland-tipped hairs that trap and digest small prey. Flowers are regular, five-petaled, and solitary, or several are clustered atop the flowering stem. There are about 200 species of sundews, and they grow on all continents except Antarctica.

Heath family (Ericaceae)
In Minnesota, this family consists of perennial herbs and shrubs. Leaves are mostly simple and alternate or opposite. Flowers are regular, generally bell- or urn-shaped, with four or five separate or fused petals. Also called the heather family, the Ericaceae contains well-known plants such as blueberries, cranberries, and rhododendrons.

Pea family (Fabaceae)
This large group of plants is also called the bean or legume family because of its leguminous fruits. Most of the species in Minnesota are annual and perennial herbaceous plants, but there are also shrubs and trees. Leaves are usually alternate and compound. Flowers are typically irregular, with five petals typically consisting of a large upper petal (the banner or standard), two lateral petals (wings), and two lower petals that are often fused (or meet at their edges), called a keel.

Gentian family (Gentianaceae)
In our area, these annual and perennial herbaceous plants have simple, entire leaves that are often opposite on the stem or whorled. Flowers are regular, four- or five-lobed, solitary or in clusters. Flowers of some gentian species possess some of the purest blue coloration in the plant world. The houseplant Persian violet (*Exacum affine*) is a member of the family.

Geranium family (Geraniaceae)

Leaves can be opposite, alternate, or basal for these annual and perennial herbs. Flowers are usually regular with five petals, in clusters atop the stem. The fruit has a long beak like a crane. The family comprises hundreds of species. Most of the cultivated geraniums (genus *Pelargonium*) have origins in South Africa.

Iris family (Iridaceae)

This is a family of perennial herbaceous plants that usually have basal, simple, sword-shaped leaves in a fanlike arrangement. Flowers are regular or nearly so, with six tepals, solitary or clustered. The commonly cultivated bearded irises of home gardens are horticultural variants involving the German iris (*Iris germanica*) of Europe.

Mint family (Lamiaceae)

Plants in this large family are often aromatic (though several species are not). Square stems are characteristic. Leaves are opposite and mostly simple. Flowers are usually irregular (sometimes nearly regular), tubular, and two-lipped with five lobes. The most aromatic species treated in this book are wild mint (*Mentha* species), wild bergamot (*Monarda* species), and mountain mint (*Pycnanthemum* species).

Bladderwort family (Lentibulariaceae)

These are carnivorous annual and perennial herbaceous plants of aquatic or saturated soils. Leaves of Minnesota species are basal and flat (*Pinguicula* species) or dissected and alternate or whorled, bearing saclike bladders (*Utricularia* species). Flowers are irregular, two-lipped with a spur at the base, and located along and atop the stem. Common butterwort (*Pinguicula vulgaris*) is restricted to areas near Lake Superior.

Lily family (Liliaceae)

This family of perennial herbaceous plants has a variety of leaf arrangements. Flowers are regular with six tepals, occurring solitary, paired, or in clusters. The lily family once consisted of several different genera, but recent taxonomic changes have resulted in many fewer. In this book, trout lilies (*Erythronium* species) and true lilies (*Lilium* species) are treated.

Loosestrife family (Lythraceae)

Most members of this family of annual and perennial herbs or shrubs grow in moist or wet habitats in Minnesota. Leaves are mostly opposite and entire. Flowers are mostly regular, with a floral cup of four to six petals that are commonly pink and crinkled.

Bunch flower family (Melanthiaceae)

Leaves of these perennial herbaceous plants range from linear to broadly elliptic and are typically mostly basal, with alternate stem leaves. Flowers are regular, with six tepals. Species were formerly included in the lily family (Liliaceae).

Miner's lettuce family (Montiaceae)

Plants in this family are somewhat succulent perennial herbs with simple, entire, and mostly opposite or basal leaves. Flowers are regular, and the Minnesota species are five-petaled. Spring beauty (*Claytonia* species) is placed in this book in the Montiaceae, but some authors place it in the purslane family (Portulacaceae).

Myrsine family (Myrsinaceae)

These mostly annual and perennial herbs have simple, opposite or whorled leaves, some of which possess visible dots or streaks. Flowers, which may also display dots or streaks, are regular, four- to nine-parted, tubular, and funnel- or bell-shaped. In our region, most species occupy wetland habitats. A few members are currently placed by some authors in the primrose family (Primulaceae).

Four-o'clock family (Nyctaginaceae)
This ornamental family of perennials is characterized by mostly simple and entire opposite leaves, the pairs unequal or equal in size. Flowers are regular, and petals are absent but with five petal-like sepals, often clumped together above the sepal-like bracts. The ornamental *Bougainvillea* species of the tropics is in this family.

Water lily family (Nymphaeaceae)
Plants of this family are distinctive, with their large, floating or emergent leaves usually notched into two lobes. The single, commonly large and showy flowers are regular, composed of numerous petals and sepals, the latter sometimes petal-like.

Evening primrose family (Onagraceae)
Plants are biennial and perennial, with opposite, alternate, or whorled leaves either entire or toothed. Flowers are often showy, normally four-parted, and regular, the floral tube elongated. Fuchsia, a mostly tropical plant often used in hanging baskets, belongs to this family.

Orchid family (Orchidaceae)
This family is often considered the largest plant family on Earth in terms of species numbers. They are perennial herbs, mostly tropical, with leaves (if present—some are leafless) being alternate, opposite, or whorled. Flowers have three petals, one of which (the lip) is typically larger, dissimilar in shape, and usually positioned lowermost.

Broomrape family (Orobanchaceae)
This large family of many genera was formerly placed in the figwort family (Scrophulariaceae). They are root parasitic

(or partially so) annuals and perennials, possessing chlorophyll or not. Leaves are simple or lobed and alternate, with some reduced to scales. Flowers are irregular, two-lipped, and five-lobed.

Wood sorrel family (Oxalidaceae)
Plants in this family are perennial herbs with mostly palmately compound leaves having three rounded leaflets notched at the tips, basal or alternate on stem. Flowers are regular, with five petals and sepals. They are known for containing oxalic acid in their tissues. The tropical starfruit (*Averrhoa carambola*) available in some supermarkets is a member of this family.

Poppy family (Papaveraceae)
This family is known for having colored or milky sap in several species, such as the red sap in bloodroot (*Sanguinaria canadensis*). Plants are herbaceous, with simple and entire or compound basal or alternate stem leaves. Flowers are regular or irregular, with most having four or more petals.

Lopseed family (Phrymaceae)
This family includes perennial herbs with simple, opposite, entire or toothed leaves. The flowers are generally irregular, tubular, and two-lipped, with the upper lip two-lobed and the lower three-lobed. Most of the members were formerly included in the figwort family (Scrophulariaceae).

Plantain family (Plantaginaceae)
This family has been revised to include species previously placed in the figwort family (Scrophulariaceae). With some reservation, I adhere to that treatment. Plants are annual or perennial herbs. Leaves are basal, alternate or opposite, entire or toothed. Flowers are regular or irregular, generally four- to five-parted.

Phlox family (Polemoniaceae)
This well-known family consists of annual and perennial herbs. Leaves are simple or compound, alternate or opposite, or basal. Flowers are regular or irregular, trumpet- or bell-shaped, and five-lobed. One species naturalized in the eastern region of Minnesota, garden phlox (*Phlox paniculata*), is widely cultivated.

Buckwheat family (Polygonaceae)
This family consists of annual and perennial herbs. Leaves are generally simple, alternate, and entire. In most of our species, papery sheaths (ocreae) wrap around swollen stem nodes. Flowers are generally small and regular, with two to six tepals, often in two whorls. Garden rhubarb (*Rheum rhabarbarum*) is a member of this family.

Buttercup family (Ranunculaceae)
This large and highly variable family comprises annual and perennial herbs with leaves both basally positioned and alternate or opposite, simple or compound. Flowers are generally regular, with three to six sepals with some petal-like; petals are absent to several. Most if not all members contain chemicals that are toxic to humans.

Rose family (Rosaceae)
Members are annual and perennial herbs, shrubs, and trees with simple or compound, mostly toothed, alternate leaves. Flowers are usually regular and five-petaled. Many well-known edible species are in the family, such as apple, blackberry, peach, and strawberry, as well as ornamentals like crabapple, rose, and spirea.

Coffee family (Rubiaceae)
Also known as bedstraw or madder family, the plants best known in the family are coffee (*Coffea* species). Ranging from

annual and perennial herbs to woody shrubs, most plants have entire, opposite leaves. Flowers are generally regular and tubular with four- or five-lobed petals.

Sandalwood family (Santalaceae)

Our single species in this family is an herbaceous perennial that is partially parasitic on roots of other plants. Leaves are mostly simple and entire, alternate or opposite. Flowers are regular, with five tepals.

Pitcher plant family (Sarraceniaceae)

This is a family of amazing carnivorous perennials that grow in wetlands. Leaves are typically modified as tubular, water-filled traps conducive to harvesting insects. Flowers are regular and nodding, with five sepals and five dangling petals surrounding a five-lobed, disklike style.

Saxifrage family (Saxifragaceae)

This family of perennial herbs has usually simple leaves in a basal circle, or if leaves are on stems, then alternate, entire or toothed. Flowers are mostly regular with five petals commonly narrowed at their bases. Many members of the family grow primarily on rocks.

Figwort family (Scrophulariaceae)

This family includes biennial and perennial herbs. Leaves are alternate or opposite, entire or toothed. Flowers are regular or irregular, with four- or five-parted petals. Many genera formerly included in this family have been assigned elsewhere, mainly to the plantain family (Plantaginaceae).

Trillium family (Trilliaceae)

This is a family of perennial herbs with solitary flowers. Its large, leafy bracts (generally viewed as leaves) occur mostly in whorls of three and are simple, entire, and ovate or obovate to elliptical. True leaves are underground, alternate, scale-like along the horizontal stem. The flower is regular with three petals. Previously, species in this family were placed in the lily family (Liliaceae).

Cattail family (Typhaceae)

Plants of this family are herbaceous perennials commonly growing in dense colonies in wetlands. Leaves are basal and alternate on the stem, two-ranked, and narrowly linear and strap-shaped, flat or triangular in cross section with sheathing bases. Flowers are minute and densely packed, without petals or sepals, the male and female blossoms separate on the same plant.

Nettle family (Urticaceae)

This family consists of annual and perennial herbs. Leaves are simple and opposite or alternate. Flowers are minute, lacking petals, with four or five sepals, with separate sexes on the same plant. Plants in this family are known for their stinging hairs, but not all species have them.

Vervain family (Verbenaceae)

These plants are annual and perennial herbs, many with four-angled stems. Leaves are simple, opposite, with margins entire, toothed, or deeply lobed or divided. Flowers are regular to irregular, tubular, with five lobes, often on long, narrow spikes.

Violet family (Violaceae)

This family consists of annuals and perennials with basal or alternate stem leaves that are simple, entire to toothed or lobed. Flowers are irregular, five-petaled, with the lowest petal often the largest and prolonged into a spur.

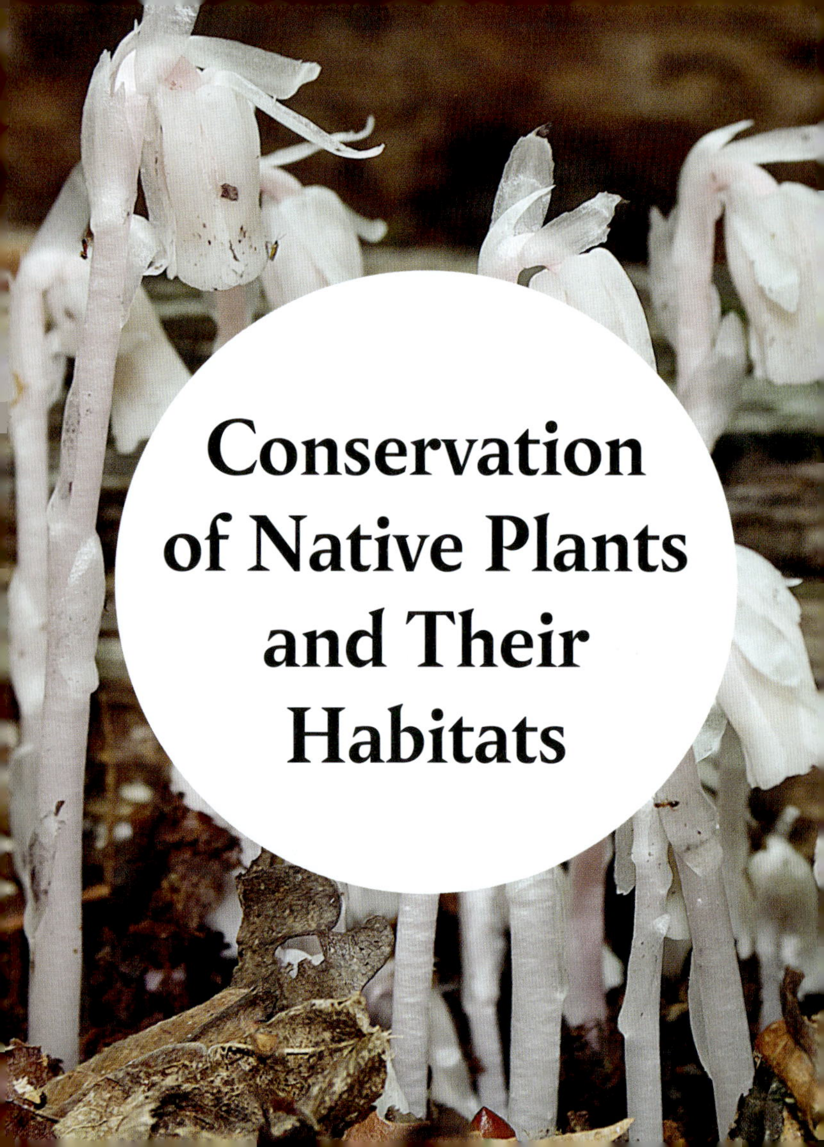
Conservation
of Native Plants
and Their
Habitats

Though wildflowers may seem abundant in places, their numbers today are only a fraction of what they once were. And the decline continues. Minnesota has a fair amount of natural land with a good diversity of wildflowers, but a constant barrage of impacts threatens to deplete it. Supporting land-protection efforts made by various public agencies and private organizations is a great way to help ensure more is not lost. If you're fortunate enough to own some land, you can try to keep it natural. But simply owning land does not guarantee protection. Factors such as the absence of fire (in fire-adapted natural communities) and the loss of large predators (especially those that prey on—and would thus keep in check—the state's overpopulation of deer) have caused a decline in our native flora. Exotic invasive plants are equally damaging. To minimize their impact, you can join efforts to control such plants, or at least avoid using invasive species in your own landscaping. You can also refrain from collecting wildflowers. Taking just one may seem harmless, but think of the impact if too many people took "just one." It's important to leave wildflowers in place for the organisms that rely on them. Native insects, especially pollinators, play an important role in the web of life. Several are generalists, being involved with a whole host of plants, but some are specialists, interacting with only one plant species. If that species disappears, so does the insect. Think about the host-specific relationship between monarchs and milkweeds—that is, monarch caterpillars feed only on the leaves of milkweeds. So if milkweeds disappear, so does this beautiful butterfly.

Using This Book

The primary intention of this book is simply to introduce readers to the wildflowers of Minnesota, and the main approach is using photographs. The photos are grouped by flower color and then alphabetically by family, genus, and species. Common names are also included. Pay close attention not only to the shape and color of the flowers but also to the arrangement of them on the plant. Also note the foliage—for example, whether the leaves are opposite or alternate—as it can be as diagnostic as the flower. Information on leaves and other characteristics described on the following pages are included to provide further help in identification.

HABITAT The environment where the plant grows. Knowing the habitat of your wildflower is quite important, as some plants are restricted to a certain habitat and are found only there. For example, some plants live only in deep water or grow only on rock.

BLOOM The season when a wildflower will most likely be in bloom. Because there is often no precise cutoff between seasons, a range may be provided, such as "summer, fall." Be aware that blooming time may fluctuate somewhat between years, as weather is often a factor.

DESCRIPTION The plant's overall growth form and stature.

FLOWER The flower's color, overall shape and size, position, and arrangement.

LEAF The location and position of leaves on the plant, plus their size and characteristics of shape, margin, and surfaces.

FRUIT The type and distinguishing features. In some cases, seeds are mentioned.

NARRATIVE Provides comparisons with similar-looking and/or related species, familiar species of the family in which it belongs, names and various uses of specific plants made by Native Americans and early colonists, origin of common and scientific names, trivia, and more. (Information regarding Native American plant use is derived from several published accounts, including interviews of tribal members made by Huron Smith of the Milwaukee Public Museum and others in the early 1900s.)

White Flowers

Sagittaria latifolia
Alismataceae

common arrowhead

HABITAT Marshes, pond and lake margins, streams, swamps, ditches

BLOOM Summer

DESCRIPTION Erect perennial herb, hairless, often in colonies, to 3 ft. tall

FLOWER White, three-petaled, female and male flower usually separate on same plant, to 1 in. wide in up to ten whorls on flowering stem

LEAF Basal, blades arrowhead-shaped, highly variable in width of lobes, toothless, hairless, on long stalks, to 1 ft. long and half as wide

FRUIT Flat nutlet with a short, bent beak

The six known species of arrowheads in the state can be difficult to differentiate based just on leaf shape. A better way is noting differences in the mature fruit. Many Indigenous Americans have used the tubers for food, as do wildlife, especially beaver and muskrat. Also known as duck potato, but its tubers are not a common food source for ducks.

Allium tricoccum
Alliaceae

wild leek

HABITAT Rich, moist forests

BLOOM Summer

DESCRIPTION Erect perennial herb, single-stemmed, hairless, to 12 in. tall, commonly forms large colonies

FLOWER White, about ¼ in. wide, with six tepals, up to 50 in a rounded cluster to 2 in. wide; when blooming, leaves are no longer present

LEAF In early spring up to three leaves, to 12 in. long and 4 in. wide, elliptic, with distinct reddish stalk and strong onion odor, dies down by late spring

FRUIT Capsule three-chambered, each bearing a shiny black seed

Known to the Ojibwe as *bûgwa'djijîca'gowûnj* (unusual onion). Also currently known as ramps in some areas, it is becoming rare from overharvesting. A closely related but less common species is the narrow-leaved wild leek (*A. burdickii*), which blooms earlier and has fewer flowers, with a leaf base that tapers gradually and lacks reddish coloration.

Cicuta maculata
Apiaceae

water hemlock

HABITAT Swamps, fens, shores, ditches, various wet sites

BLOOM Summer

DESCRIPTION Branched biennial/perennial herb with green, purple, or purplish blotched hollow stems, hairless, to 6 ft. tall

FLOWER White, five-petaled, about ⅛ in. wide, numerous in rounded, somewhat flat-topped clusters at stem tips

LEAF Compound, with leaflets to 4 in. long, margins sharply toothed, veins ending in notches between teeth, a trait not found in most plants

FRUIT Roundish, ribbed, each containing two seeds

Water hemlock is perhaps the most poisonous native plant of North America. It is reported that if one eats it, especially the roots, death can occur within a matter of minutes. It is related to poison hemlock (*Conium maculatum*), an introduced species from Europe found here mostly in disturbed areas in the southern part of the state. Poison hemlock was used to put Socrates to death.

Cryptotaenia canadensis
Apiaceae

honewort

HABITAT Moist forests, stream terraces, ravines

BLOOM Summer

DESCRIPTION Erect, branching perennial, hairless, to 3 ft. tall

FLOWER White, five-petaled, to 1⁄16 in. wide, in spreading clusters at stem tips

LEAF Alternate, compound with three leaflets, to 6 in. long and slightly more than half as wide, irregularly toothed

FRUIT Narrow, ribbed, and two-seeded, resembles fennel seeds

A member of the carrot family, honewort is one of the few forest plants to bloom during summer. The flowers are quite tiny. A close relative, mitsuba (*C. japonica*), is used in salads and soups in Japan. Honewort was used to treat hone, or swelling of the cheek.

Heracleum maximum
Apiaceae

cow parsnip

HABITAT Wet meadows, thickets, floodplains, streambanks

BLOOM Spring, summer

DESCRIPTION Large, upright perennial herb, hairy, stems strongly ridged and hollow, to 8 ft. tall

FLOWER White, five-petaled, each with notched tip, the outer petals larger than the inner ones, to ¼ in. wide, in flat-topped, grouped clusters, to 10 in. wide

LEAF Compound with three leaflets, to 15 in. long and about as wide, lobed and toothed, alternate with a papery sheath clasping the main stem

FRUIT Flattened, somewhat oval, about ¼ in. wide, containing a pair of seeds

Leaves and flower heads of this impressively large plant are hard to miss. One source reports it as one of the top-ten plant species used for a variety of reasons by Indigenous Americans. It also contains compounds that, when in contact with skin and exposed to light, may cause severe dermatitis.

Osmorhiza longistylis
Apiaceae

aniseroot

HABITAT Rich, moist forests

BLOOM Spring

DESCRIPTION Erect perennial herb, branched, hairless or hairy, stem commonly reddish in color, to 3 ft. tall

FLOWER White, five-petaled, to ⅛ in. wide, styles longer than the petals, up to 50 in flat-topped clusters at stem tips

LEAF Fernlike, subdivided into leaflets that create an overall triangular outline, margins toothed, alternate, hairless or somewhat hairy, produces aroma of anise when crushed, to 9 in. long

FRUIT Narrow, slightly curved, tapered at both ends, to about 1 in. long, with persistent styles to ⅛ in. long

Aniseroot is recognized in part by its flat-topped cluster of flowers, but also by its roots and leaves that smell like anise. The Ojibwe have used it to make tea to treat sore throats. The similar-looking sweet cicely (*O. claytonii*) differs by having little or no anise scent and shorter styles.

Apocynum androsaemifolium
Apocynaceae

spreading dogbane

HABITAT Dry savannas, open woodlands, prairies, fields, roadsides

BLOOM Summer

DESCRIPTION Perennial herb with spreading branches, stems smooth and commonly reddish, milky sap, to 3 ft. tall

FLOWER White to pinkish exterior, interior with pink stripes, bell-shaped with five recurved lobes, nodding, assembled in clusters at branch tips

LEAF Oval, unlobed, oppositely attached to stem, to 4 in. long and half as wide, often drooping

FRUIT Resembles skinny, dangling string beans, to 6 in. long

Compared to spreading dogbane, common dogbane (*A. cannabinum*) is usually taller and has all-white flowers. Their stems are a source of fiber that can be used to make cordage (rope). Its outer rind has been used as thread for fine sewing by the Ojibwe. Dogbanes are a food source for the beautiful dogbane beetle (*Chrysochus auratus*).

Calla palustris
Araceae

wild calla

HABITAT Shallow, mostly acidic water in bogs, swamps, lake margins, slow-moving streams, ditches

BLOOM Spring

DESCRIPTION Colony-forming perennial, hairless, to 15 in. tall

FLOWER Greenish white, tiny, several on a thick cylindrical spike that emerges from base of a broad, white, petal-like spathe (blade), to 4 in. long and 2 in. wide

LEAF Basal, heart-shaped, shiny, with prominent midvein and numerous side veins, to 5 in. long

FRUIT Cluster of bright red (when ripe) berries

Different from the similar-looking calla lily (*Zantedeschia* species) of the florist trade. Neither species is a member of the lily family (Liliaceae), though the two are related—both occur in the arum family (Araceae) along with other Minnesota natives such as jack-in-the-pulpit and skunk cabbage.

Aralia nudicaulis
Araliaceae

wild sarsaparilla

HABITAT Moist forests, woodlands, swamps, dunes

BLOOM Spring, summer

DESCRIPTION Erect, three-pronged perennial, to 2 ft. tall, colony-forming

FLOWER White or greenish, five-petaled, to ⅛ in. wide, numerous, usually in three rounded clusters, each to 2 in. wide atop a naked (leafless) stem

LEAF Pinnately compound, with normally five leaflets, these to 5 in. long and 2 in. wide, finely toothed, copper to bronze color and shiny when young

FRUIT Bluish black berry to ¼ in. wide

Named for its use as a substitute for sarsaparilla. Its berries are eaten by several forest birds including White-throated Sparrow and Ruffed Grouse and are an important food for black bears during summer and fall foraging. The plant has been used medicinally for a variety of ailments and is related to ginseng (*Panax quinquefolius*), a popular medicinal herb.

Achillea millefolium
Asteraceae

yarrow

HABITAT Fields, roadsides, grasslands, woodlands

BLOOM Summer

DESCRIPTION Erect perennial herb, usually unbranched except at top, somewhat woolly, often clumping, to 2½ ft. tall

FLOWER White, also pink or red, in compact heads to ½ in. wide, composed of outer ring of five white ray flowers and a central disk of up to 20 tiny tubular flowers, arranged in flat-topped clusters atop the stem

LEAF Finely dissected, feathery or fernlike, to 6 in. long and 1 in. wide, alternate, aromatic when bruised

FRUIT Small, dry one-seeded nutlet

Presumably native, but some plants may be introduced from Eurasia (where the species is also native). The genus *Achillea* is named for the Greek god Achilles, who is said to have used yarrow to treat his wounds. For North American Indigenous people, yarrow is reported to have had the greatest number of medicinal uses than any other plant.

Ageratina altissima
Asteraceae

white snakeroot

HABITAT Moist to dryish forests, open woodlands, thickets

BLOOM Late summer, fall

DESCRIPTION Erect perennial, stems mostly smooth and hairless, branching in flower clusters, to 3 ft. tall

FLOWER White, tiny and tubular, up to 20 in. heads about ¼ in. wide, occurring in somewhat flat-topped clusters

LEAF Triangular with sharply toothed margins, opposite, to 6 in. long and 4 in. wide

FRUIT Slender nutlet with tuft of hairs at one end that enables it to be carried in the wind

This species is attributed to having killed Nancy Hanks Lincoln, Abraham Lincoln's mother. It contains a toxic chemical that when eaten by a cow is transferred to her milk and to the person who drinks it, causing "milk sickness." It was a significant cause of death for many pioneers. *Eupatorium rugosum* is its former scientific name.

Antennaria neglecta
Asteraceae

field pussytoes

HABITAT Prairies, fields, open woods, roadsides

BLOOM Spring

DESCRIPTION White-woolly, matt-forming perennial herb with erect flower stalks to 1 ft. tall, spreads by runners

FLOWER White, tiny, male and female on separate plants, several in heads to ⅓ in. wide

LEAF Mostly basal, rather narrowly spoon-shaped with a single vein running through the middle, white-woolly beneath

FRUIT Seedlike fruit bears a fluffy tuft of hairs that facilitates wind dispersal

This is one of six species of pussytoes in the state, the field pussytoes being the most widely distributed. The common name reflects the appearance of the furry flowering heads, somewhat like the hairy pads of a cat's foot. The genus name alludes to the male flower's club-tipped bristles, said to resemble insect antennae.

Erigeron philadelphicus
Asteraceae

Philadelphia fleabane

HABITAT Moist forest clearings and edges, streambanks, roadsides, trails, meadows, fields

BLOOM Spring, summer

DESCRIPTION Erect, hairy biennial or perennial herb, branched above, to 3 ft. tall

FLOWER White to pinkish, in compact heads to ¾ in. wide, with an outer ring of 100 or more threadlike ray flowers and a central disk of equal number of tiny yellow tubular flowers, in spreading clusters atop the stem

LEAF Basal leaves coarsely toothed, stem leaves usually less so or not at all, bases usually clasping the stem, alternate, to 3 in. long and 1 in. wide

FRUIT Dry, about 1⁄16 in. long, with tuft of long hairs that facilitates wind dispersal

Fleabanes reportedly are used to eradicate or repel fleas, though no evidence proves that they do either. Compared to other fleabanes in the state, this species differs by having clasping leaves. Its Ojibwe name is *mîcaoǵgacan* (odor of deer hooves). Smoke from its burning flowers is said to attract bucks.

garlic mustard

Berteroa incana
Brassicaceae

hoary alyssum

HABITAT Fields, pastures, roadsides, disturbed areas, especially in sandy soil

BLOOM Summer, fall

DESCRIPTION Erect, grayish green, densely hairy annual or short-lived perennial herb, branched toward tip, to 3 ft. tall

FLOWER White, about ¼ in. wide, with four deeply lobed petals (appears as eight), in rounded clusters about 2 in. wide at stem tips

LEAF Basal (when young, absent as plant matures) and alternate on stem, hairy, unlobed, toothless

FRUIT Elliptic, somewhat flattened with a skinny beak

Hoary alyssum is reported to be the most common poisonous plant to horses in Minnesota. It blooms throughout the growing season. Another white-flowered invasive plant of the mustard family is garlic mustard (*Alliaria petiolata*, see inset), which is mostly a forest plant, blooming only in spring. It has triangular-shaped leaves that smell of garlic when crushed.

Capsella bursa-pastoris
Brassicaceae

shepherd's purse

HABITAT Disturbed areas, roadsides, vacant lots, agricultural fields

BLOOM Spring, summer, fall

DESCRIPTION Somewhat-branched annual to 2 ft. tall, with simple and star-shaped hairs toward its base

FLOWER White, four-petaled, to ⅛ in. wide, clustered at stem tips; one of the earliest plants to bloom

LEAF Basal rosette, hairy, deeply lobed, stem leaves lance- or arrowhead-shaped, alternately attached with clasping bases

FRUIT Triangular to sharply heart-shaped, the narrow end at base, presenting a distinctive purselike appearance

This native of Eurasia is now a worldwide weed. When wet, seeds produce a toxic, mucilaginous coating that attracts and kills various organisms such as nematodes. Compounds from the decaying nematodes are reported to benefit the plant's growth, and thus some botanists suggest shepherd's purse could be categorized as a carnivorous plant.

Cardamine concatenata
Brassicaceae

cutleaf toothwort

HABITAT Rich, moist forests, ravines, slopes, stream terraces

BLOOM Spring

DESCRIPTION Erect perennial herb, unbranched, mostly hairless, to 12 in. tall

FLOWER White, four-petaled, to ½ in. wide, in loose cluster atop the stem

LEAF Basal, on stem usually three in a whorl, three- to five-parted, toothed, to 5 in. long and about as wide or wider

FRUIT Narrow, upright pods splitting lengthwise in halves

Toothwort literally means "tooth plant" for the toothlike bumps on the underground rhizome of some closely related species (not so much on this species). The rhizomes are said to have a peppery taste, somewhat like radish. Not surprisingly, both toothwort and radish are in the mustard family. Known in eastern half of state only.

Silene latifolia
Caryophyllaceae

white campion

HABITAT Disturbed sites, roadsides, railroads, pastures, ditches, waste places

BLOOM Summer, fall

DESCRIPTION Erect or ascending annual or short-lived perennial, branched above with sticky, glandular-tipped hairs when mature, to 3 ft. tall

FLOWER White, to 1 in. wide, with five, two-lobed petals attached at tip of inflated calyx (when mature); male and female flowers on separate plants; calyx of male plants have 10 reddish, hairy veins, 20 in females, at tips of spreading branches, opening in evening

LEAF Lance-shaped to elliptic, unlobed, glandular hairy, opposite, to 5 in. long and 1¼ in. wide, smaller leaves commonly in axils of main leaves

FRUIT Tan capsule with five spreading, split teeth (looks like ten) at tip

This Eurasian species belongs to a group of plants known as catchflies. Their sticky glands often trap tiny insects, leading some to consider this a primitive form of carnivory. One source says *Silene* is from *sialon* (Greek for "saliva"), referring to the plants' sticky secretions.

Maianthemum canadense
Convallariaceae

Canada mayflower

HABITAT	Dry to moist woods, even in dense shade, swamps, bogs
BLOOM	Spring, summer
DESCRIPTION	Erect perennial herb with a somewhat zigzagging stem, hairy or not, to 8 in. tall
FLOWER	White, four tepals, spreading or recurved, to ¼ in. wide, scattered on stalk atop the plant
LEAF	Heart-shaped, alternate, clasping, two or three per plant, hairy or not, to 3 in. long and 2 in. wide
FRUIT	Berry, when ripe, is red and speckled, about ¼ in. wide

The genus name is from the Latin *Maius* (May) and the Greek *ánthemon* (flower), giving us the common name mayflower. It is sometimes called wild lily-of-the-valley for its superficial resemblance to the garden plant *Convallaria majalis*. The Ojibwe name, *agoñgosi'mînûn*, means "chipmunk berries." One use was to cure headaches.

Maianthemum racemosum
Convallariaceae

Solomon's plume

HABITAT Moist to dryish woods, stream terraces, thickets

BLOOM Spring, summer

DESCRIPTION Arching perennial herb, with somewhat zigzagging stem, unbranched, hairless to finely hairy, to 2½ ft. tall

FLOWER White, with six spreading tepals and six conspicuous stamens longer than the tepals, to ¼ in. wide, in a feathery, somewhat triangular-shaped cluster of flowers at the stem tip

LEAF Elliptic, noticeably parallel veined, somewhat on same plane, stalkless, alternate, to 6 in. long and 3 in. wide

FRUIT Round berry, red when ripe

Because Solomon's plume's foliage resembles that of smooth Solomon's seal, another name is false Solomon's seal. Also in Minnesota is starry Solomon's plume (*M. stellatum*), with tepals that are longer than the stamens and narrower and slightly clasping leaves.

Calystegia sepium
Convolvulaceae

hedge bindweed

HABITAT Thickets, streambanks, fencerows, ditches, wooded edges, weedy places

BLOOM Summer, fall

DESCRIPTION Twining, hairless perennial vine without tendrils, to 10 ft. long

FLOWER White, pink, or bicolored, funnel-shaped, shallowly five-lobed, to 3 in. wide at mouth, with two broad green bracts at the flower base

LEAF Alternate, arrowhead-shaped, with angular lobes at base, hairless, to 4 in. long

FRUIT Rounded capsule, about ½ in. wide

This species is similar-looking and related to morning glory (*Ipomoea* species), but hedge bindweed flowers differ by having two large bracts at their bases. Interesting common names include old man's nightcap, wedding gown, and belle of the ball. The Latin name *sepium* means "of hedges." The flower color is quite varied. Plants are poisonous.

Cuscuta gronovii
Convolvulaceae

common dodder

HABITAT Grows on other plants usually occupying wet sites such as swamps, bottomland forests, fens, streambanks

BLOOM Summer

DESCRIPTION Twining annual vine, yellowish orange, branching, hairless, to 6 ft. or longer

FLOWER White, bell-shaped, with five spreading, slightly rounded lobes, to ⅛ in. wide, in clusters on side branches

LEAF Minute, scalelike if present, alternate, and same color as stem

FRUIT Rounded capsule, somewhat greenish and swollen at tip

These curious plants are parasites that use small suckerlike structures to withdraw water and nutrients from stems of jewelweeds, asters, and other plants. They resemble strands of cooked spaghetti draped over their hosts. Fortunately, they rarely kill the host. This one is the most widespread of the species in the state, among several others, including some that are quite rare.

peeled dried fruit

Echinocystis lobata
Cucurbitaceae

wild cucumber

HABITAT Floodplain forests, thickets, forest edges, fencerows

BLOOM Summer, fall

DESCRIPTION Climbing annual vine with coiling tendrils, stems angled and hairless, to 10 ft. long

FLOWER White, somewhat cup-shaped with six narrow, strap-shaped petals, male and female flowers separate on same plants, males several in elongated clusters, females fewer on shorter stalks, both to ½ in. wide

LEAF Alternate, with five triangular lobes radiating like fingers on a hand, hairless, to 4 in. long and about as wide

FRUIT Oblong, smooth-skinned fruit with soft prickles, to 2½ in. long

Also known as balsam apple, though its fruit is not considered edible. When thoroughly dried and peeled, fruit resembles a miniature luffa sponge. The genus name is from the Greek *echinos* (hedgehog or sea urchin) and *cystis* (bladder), a reference to its inflated spiny fruit. The Ojibwe have made tea from its roots to treat stomach troubles.

sundew flower

Drosera rotundifolia
Droseraceae

round-leaved sundew

HABITAT Sphagnum bogs, fens, marshes, shores, wet sand prairies

BLOOM Summer

DESCRIPTION Small carnivorous herb to about 2 in. wide, essentially stemless except for flowering stalk, to 10 in. tall

FLOWER White, five-petaled, to ⅓ in. wide, blooming in ascending order along the side of an unfurling leafless stem

LEAF Basal rosette, often reddish, modified into round pads at ends of stalks, covered with hairs tipped with sticky droplets

FRUIT Egg-shaped capsule

There are four species of sundew in Minnesota. All have sticky hairs that trap small insects that slowly dissolve to provide nutrients to the plant. This species occurs in northern latitudes around the world. Darwin wrote of it, “Is it not curious that a plant should be far more sensitive to a touch than any nerve in the human body!”

Monotropa uniflora
Ericaceae

Indian pipe

HABITAT Moist forests, usually in acidic soils

BLOOM Summer, fall

DESCRIPTION Erect perennial herb, translucent white to pinkish, fleshy, unbranched, hairless, to 8 in. tall

FLOWER White, bell-shaped, to ¾ in. long, with three to six petals nodding then turning vertical with age, occurring singly atop stem

LEAF Reduced to scales, alternate, white, to ⅔ in. long and ¼ in. wide

FRUIT When mature an erect, black, round capsule

This amazing plant lacks chlorophyll and the ability to photosynthesize. It survives because its roots connect to fungi that supply it with food drawn from roots of green plants. Also known as ghost flower because of its pale white appearance. It is called *xawiska* (flowers white) by the Ho-Chunk people, who have used it to revive someone who has fainted.

Pyrola elliptica
Ericaceae

large-leaved shinleaf

HABITAT Moist to dry forests, woodland slopes, in acidic soil

BLOOM Summer

DESCRIPTION Erect perennial herb, unbranched, hairless, grows low to the ground, to 1 ft. tall when flowering

FLOWER White, about ½ in. wide, with five oval-shaped, nodding petals bearing green veins, and a downwardly pointing style with an upcurved tip, several positioned alternately on the flowering stalk to 1 ft. tall

LEAF Basal, elliptic, thin, usually not very shiny, hairless, margins often with minute teeth, to 3 in. long and half as wide

FRUIT Dry, round capsule

Minnesota has five species of *Pyrola*, with the large-leaved shinleaf apparently the most common. The whole group is often referred to as shinleaf. It is reported that their leaves contain an aspirin-like substance that will relieve pain when ground up and applied as a paste to shins (and, it would seem, to other body parts).

ground fruit pod

Amphicarpaea bracteata
Fabaceae

American hog peanut

HABITAT Moist to dryish forests, thickets, seeps

BLOOM Summer, early fall

DESCRIPTION Twining herbaceous annual vine, variously hairy, to 6 ft. long; two varieties, one with purple flowers and spreading stem hairs (var. *comosa*), the other with white flowers and few hairs (var. *bracteata*), with some overlap in features between the two

FLOWER White or purple, typical pea family shape, to ¾ in. long, in clusters on long stalks extending from leaf axils

LEAF Alternate, divided into three leaflets, oval or diamond-shaped, hairy or smooth, to 4 in. long and 3 in. wide

FRUIT Pod, similar to a garden pea, with edible seeds; or a round pod that grows at ground level

Produces two types of flowers, one as pictured in the photo, the other lacking petals. The latter type forms an edible, round "peanut" at ground level or underground. The Ojibwe have used the seeds from the aboveground pods for food.

Glycyrrhiza lepidota
Fabaceae

wild licorice

HABITAT	Moist prairies, fields, roadsides, disturbed sites
BLOOM	Summer
DESCRIPTION	Erect perennial herb, hairless or sparsely hairy but covered with black dots toward the tip, colonial, to 3 ft. tall
FLOWER	White or greenish white, somewhat tubular, about ½ in. long, five-petaled with pea flower shape, the uppermost petal larger, occurring in dense clusters from atop stem and in axils of upper leaves
LEAF	Pinnately compound, alternate, to 5 in. long, with 11–19 lance-shaped leaflets
FRUIT	Cylindrical pod covered with hooked prickles, reminiscent of a small cocklebur, about ½ in. long

The root is said to be sweet, due to the presence of glycyrrhizin, a substance also found in the related and similar licorice plant of Europe (*Glycyrrhiza glabra*) from which licorice extract is derived. The burlike fruits attach to fur and clothing, which helps them disperse to other locations where new populations can be established.

Trifolium repens
Fabaceae

white clover

HABITAT Lawns, pastures, roadsides, creekbanks, gardens, disturbed places

BLOOM Spring, summer, fall

DESCRIPTION Perennial herb, grows along the ground via trailing stems that root at nodes (where leaves are attached), unbranched, mostly hairless, to 10 in. tall

FLOWER White, tubular, with a typical pea flower shape, five-petaled, several in a rounded head, to ¾ in. wide, positioned on a single leafless stalk

LEAF Compound, alternate, with three broadly elliptic leaflets, unlobed, margins finely toothed, often with a whitish chevron patch, to 1 in. long and about as wide

FRUIT Small pod hidden within the withered flower

The genus name of this Eurasian native reflects the three leaflets—*tri* (three) and *folium* (leaf). Many of the attributes of red clover are also present in white clover, including nitrogen fixation and use as a forage crop. White clover is the most likely clover to be growing in lawns.

Mentha canadensis
Lamiaceae

wild mint

HABITAT Marshes, swamps, fens, meadows, shores, ditches

BLOOM Summer, fall

DESCRIPTION Erect to sprawling perennial herb, square-stemmed, usually unbranched, variously hairy (sometimes mostly on the angles), with a strong minty scent when bruised, rhizomatous, to 2 ft. tall

FLOWER White to pale purple, tubular, with two flaring lips, the upper one with two lobes, the lower one with three, to ⅛ in. long, densely packed in whorls around the leaf axils

LEAF Elliptic with sharp teeth, opposite, pairs at 90 degrees to one another, to 2½ in. long and ¾ in. wide

FRUIT Small, dry nutlets

Also known as *M. arvensis*, this is our only native *Mentha* mint. Several other *Mentha* species are also found here, introduced mostly from Europe. Water horehounds (*Lycopus* species) look similar but lack a minty scent. Wild mint has often been used by Native Americans for medicinal purposes. It is a natural source of menthol.

Pycnanthemum virginianum
Lamiaceae

common mountain mint

HABITAT Prairies, marshes, fens, meadows, fields

BLOOM Summer, fall

DESCRIPTION Erect perennial herb, branched on upper plant, four-angled with hairs on the angles, often colonial, to 3 ft. tall

FLOWER White, to ⅛ in. wide, tubular, two-lipped, purple-spotted, upper lip shallowly notched, lower lip three-lobed, in flat heads to ¾ in. wide

LEAF Narrowly lance-shaped, tapering to a point, hairless, unlobed, toothless, and stalkless, opposite, very aromatic when crushed, to 2½ in. long and ½ in. wide

FRUIT Dry, capsule-like with four nutlets

Pycnanthemum is from the Greek *pyknos*, meaning "dense," and *anthemon* for "flower," referring to its densely packed flower heads. Insects, especially butterflies, are strongly attracted to mountain mint flowers. It is sometimes used as an ornamental, perhaps because it is said to be deer-proof.

Erythronium albidum
Liliaceae

white trout lily

HABITAT Moist woods, floodplains, slopes, ravines

BLOOM Spring

DESCRIPTION Erect to reclining perennial herb, unbranched, hairless, colonial by stolons, to 7 in. tall

FLOWER White, but may be pinkish tinged, bell-shaped, about 1½ in. long, with six reflexed tepals occurring singly atop the flowering stem

LEAF Elliptic to lance-shaped, somewhat pliable and waxy-looking, pale or mint-green, usually mottled with brown or purplish splotches; flowering plants have two leaves, nonflowering (the majority of plants) only one

FRUIT Oval capsule, usually held erect

This species occurs primarily in the eastern half of the state. The dwarf trout lily (*E. propullans*) also has white flowers, occurring in far southeastern Minnesota and nowhere else on Earth. Its flowers are considerably smaller than white trout lily and often have only four or five tepals. Yellow trout lily (*E. americanum*) occurs in the far eastern parts of the state.

Anticlea elegans
Melanthiaceae

white camas

HABITAT Moist to wet prairies, meadows, grassy hillsides

BLOOM Summer

DESCRIPTION Erect perennial herb, unbranched, hairless, from a bulb, to 3 ft. tall

FLOWER White, with six spreading tepals, each bearing a yellowish green gland, to ¾ in. wide, several in a spreading cluster atop the plant

LEAF Linear, mostly basal, those on stem much reduced, unlobed, hairless, to 1 ft. long and ½ in. wide

FRUIT Three-lobed, cone-shaped capsule

Also known as mountain death camas, this plant is deadly poisonous. Its pollen and nectar can even kill insects that feed on them. The "camas" portion of its name comes from its similarity to camas (*Camassia quamash*), an unrelated western plant with edible bulbs. *Anticlea* is named for the mother of the mythical Odysseus.

Claytonia virginica
Montiaceae

Virginia spring beauty

HABITAT Moist to dryish forests

BLOOM Spring

DESCRIPTION Erect to ascending perennial herb, unbranched, hairless, to 6 in. tall (usually much less)

FLOWER White, about ½ in. wide, five-petaled, commonly with pink stripes, loosely arranged atop an elongating flowering stalk

LEAF Basal, plus a single pair oppositely arranged on flowering stem, linear to narrowly elliptic, somewhat fleshy, stalks indistinct, to 4 in. long and ½ in. wide

FRUIT Round capsule cradled in a cuplike pair of persistent sepals

Spring beauty belongs to a class of plants called spring ephemerals. They arise from the ground in early spring before the leaves of trees appear. Their tops die back soon after the tree leaves fully develop. Carolina spring beauty (*C. caroliniana*) occurs in the far northeastern counties. Each leaf has a distinct stalk.

Nymphaea odorata
Nymphaeaceae

fragrant white waterlily

HABITAT Ponds, lakes, calm water portions of rivers, often rooted in highly organic soil

BLOOM Summer, fall

DESCRIPTION Aquatic perennial herb with branching rhizomes from which the leaves grow, the stalks of the latter to 6 ft. long

FLOWER White, solitary on long stalks, many-petaled, floating to slightly perched above the water surface, fragrant, to 7 in. wide

LEAF Mostly round with a slit on one side from which most main veins spread palmately, shiny, often purple underneath, to 1 ft. long and wide

FRUIT Berrylike, fleshy, matures underwater on coiled stalks

The name *Nymphaea* comes from the mythological water nymph, a female deity that lives in watery places, the same environment where waterlilies occur. With so many natural lakes, Minnesota is the perfect place for waterlilies. Flowers open in the morning and close in the evening. (The smaller leaves pictured here are watershield.)

Circaea canadensis
Onagraceae

enchanter's nightshade

HABITAT Moist forests, ravines, stream terraces, thickets

BLOOM Summer

DESCRIPTION Relatively small, erect perennial herb, sparsely hairy on at least upper stem and branches, to 2 ft. tall

FLOWER White, approximately ⅛ in. wide, with two petals that are notched, giving the impression of four petals

LEAF Opposite pairs, these at right angles to adjacent leaf pairs, lance-shaped, mostly hairless, with scattered teeth along margin, to 4 in. long and about half as wide

FRUIT Capsule, grooved on the surface, bearing hooked hairs that enable it to attach to clothing and fur

Fruits of this plant stick tight to clothing or fur via tiny hooks that grab onto passersby. The similar alpine enchanter's nightshade (*C. alpina*) is more delicate and lacks grooves on the fruit. These two are not true nightshades or at least are not members of the nightshade family (Solanaceae).

squirrel corn

Dicentra cucullaria
Papaveraceae

Dutchman's breeches

HABITAT Rich, moist forests, wooded ravines, stream terraces

BLOOM Spring

DESCRIPTION Erect perennial herb, unbranched and hairless, with pinkish white underground bulblets, to 1 ft. tall

FLOWER White, to ¾ in. long, with two inconspicuous petals and two large, inflated, upwardly spreading petals that resemble pointed spurs (the "legs"), dangling upside down on an arching stem, like miniature trousers on a clothesline

LEAF Basal, highly dissected into narrow lobes, giving a feathery appearance, hairless, grayish green, pale beneath

FRUIT Narrow capsules that taper at each end, resemble tiny pea pods

Flowers of Dutchman's breeches are similar in appearance to the less common squirrel corn (*D. canadensis*, see inset), but the latter's two larger petals are not spreading, its underground bulblets are yellow, and it is confined to the far southeastern corner of the state. The Asian bleeding-heart (*Dicentra spectabilis*) is an ornamental relative.

Sanguinaria canadensis
Papaveraceae

bloodroot

HABITAT Rich, moist, forested ravines, stream terraces, slopes

BLOOM Spring

DESCRIPTION Erect perennial herb, unbranched, hairless, from a rhizome with blood-red sap, to 10 in. tall

FLOWER White, with a ring of up to 12 elliptic petals surrounding numerous yellow stamens, solitary, to 2 in. wide

LEAF Single, rounded in outline with up to seven deep lobes at the tip, rubbery in texture, hairless, to 6 in. long and about as wide

FRUIT Capsule; seeds with gelatinous food body attached

Named for its blood-red sap, *Sanguinaria* is from the Latin *sanguis* (blood) and *sanguinarius* (bleeding). Each seed has a food body attached that attracts ants. The ants take the seeds to their nests, consume the gelatinous food bodies, and discard the seeds, leaving them to germinate and potentially form new plants.

Phryma leptostachya
Phrymaceae

American lopseed

HABITAT Moist to dryish forests, ravines, stream terraces

BLOOM Summer

DESCRIPTION Erect perennial herb, mostly unbranched, hairy, stems four-angled, to 2 ft. tall

FLOWER White, tubular, about ¼ in. long, two-lipped, the small upper lip pinkish purple, notched in the middle and curved upward, the lower lip three-lobed, occurring in pairs evenly spaced on stalks atop plant and in leaf axils, initially held horizontally, drooping with age

LEAF Oval, toothed, opposite, pairs positioned at 90 degrees to one another

FRUIT Dry nutlet with hooks at the tip, pressed down against the stem

Flowers are often overlooked because of their small size. Its distinctive fruit is tightly appressed downward (like ears of a lop-eared rabbit), thus the common name. Each fruit has three hooks at one end, suggesting a mechanism for attaching to fur or clothing, but rarely seems to do so.

Chelone glabra
Plantaginaceae

white turtlehead

HABITAT Marshes, fens, stream borders, seeps, forested swamps

BLOOM Late summer, fall

DESCRIPTION Erect perennial herb, hairless, with square stem, to 4 ft. tall

FLOWER White, some with pink-tinged tips, two-lipped, tubular, several in cluster at stem tip

LEAF Opposite, narrow, spear-shaped with toothed margins, to 6 in. long and 1½ in. wide

FRUIT Round to oval capsule, brown when mature

Both common and Latin names reference the flower shape, like that of a turtle's head. *Chelone* is Greek for "tortoise." Although typically all white, some flowers may have a blush of pink at the tips. White turtlehead occurs mostly in the eastern half of the state.

Veronicastrum virginicum
Plantaginaceae

Culver's root

HABITAT Prairies, savannas, open woodlands, meadows, fields

BLOOM Summer

DESCRIPTION Erect perennial herb, unbranched, mostly hairless, to 6 ft. tall

FLOWER White, tubular to ⅓ in. long, with four short lobes and long exerted pollen-bearing stamens, numerous in vertical, candelabra-like spikes to 8 in. long

LEAF Lance-shaped in whorls of four or five, toothed, to 5 in. long and 1 in. wide

FRUIT Elliptical to egg-shaped capsules, seeds numerous

Native Americans and early colonists used the plant to treat tuberculosis. Referred to as "Culver's root" in a letter dated 1716 by Puritan Cotton Mather to John Winthrop, it is likely named for an acquaintance of Mather's identified in the letter as "Mr. Culver" from Lebanon, Connecticut. Why it's named for him is unclear.

Actaea rubra
Ranunculaceae

red baneberry

HABITAT Moist upland forests and woodlands, forested swamps

BLOOM Spring

DESCRIPTION Erect, branched perennial, to 3 ft. tall

FLOWER White, about ¼ in. wide, four to eight petals, short-lived, persistent stamens white and showy, numerous in a rounded cluster to 4 in. long, at tip of stalk well above foliage

LEAF Compound, two or three times divided, leaflets about 2 in. long and wide, coarsely toothed, undersurface somewhat hairy on veins

FRUIT Several seeded, shiny red (rarely white) berries, on slender green stalk

Poisonous to humans, especially its berries, but the latter are eaten by birds and small mammals with no apparent harm. The similar white baneberry, or dolls eyes (*A. pachypoda*), occurs in far eastern Minnesota. Its berries are typically white, rarely red, and the stalk of each is usually much thicker than that of red baneberry.

Anemone canadensis
Ranunculaceae

Canada anemone

HABITAT Moist sedge meadows, wet prairies, shorelines, floodplains, road banks

BLOOM Spring, summer

DESCRIPTION Short-hairy, rhizomatous perennial herb, to 2 ft. tall, colonial

FLOWER White, to 1½ in. wide, five sepals are petal-like (true petals absent)

LEAF Basal and leaflike bracts deeply three- to five-lobed and toothed, basal leaves on long stalks, bracts opposite and stalkless, all to 6 in. across

FRUIT Flattened seed with a noticeable beak, several packed into a tight cluster

This attractive wildflower, also known as meadow anemone, has tempted many to consider using it in their home landscaping, but it can aggressively spread. The Ojibwe name is *mîdewidji'bîk* (medicine lodge root), as its roots have been used as throat lozenges to help one sing better in the medicine lodge ceremony.

false rue anemone
rue anemone

Anemone quinquefolia
Ranunculaceae

wood anemone

HABITAT Moist to relatively dry forests, swamps, thickets, streambanks

BLOOM Spring

DESCRIPTION Erect, low-growing perennial herb, hairy, unbranched, colonial by rhizomes, to 8 in. tall

FLOWER White, usually with five elliptic, petal-like sepals, to 1 in. wide, single on a slender stalk positioned above the leaves

LEAF Usually in a single whorl of three, each with mostly three lobes, but the outer two segments can be split, giving the overall impression of five pointed segments per leaf, hairy, to 2 in. long and about half as wide

FRUIT Oval with a noticeable beak, several packed into a rounded cluster

The epithet means five-leaved (but see leaf description). Other white-flowering forest plants in southeastern Minnesota that bloom about the same time are false rue anemone (*Enemion biternatum*), with five petal-like sepals, and rue anemone (*Thalictrum thalictroides*), which usually has more. Both have hairless leaflets with somewhat rounded bases and tips.

Clematis virginiana
Ranunculaceae

virgin's bower

HABITAT Forest edges, thickets, fencerows, streambanks

BLOOM Summer

DESCRIPTION Climbing perennial vine to 15 ft. long, branching, with sparsely hairy and angled stems

FLOWER White, with four petal-like sepals, about ¾ in. wide, male and female flowers on separate plants

LEAF Compound, in opposite pairs on stem, three leaflets, sparsely hairy and coarsely toothed, to 3 in. long and about as wide

FRUIT Seedlike with an attached feathery plume, clustered to form a rounded, fuzzy ball

Virgin's bower is related to the commonly cultivated ornamental clematis. Two native species are in Minnesota, this one and purple clematis (*C. occidentalis*). The latter is less common and mostly restricted to the northeastern counties. As its name implies, its flowers are purple and are also larger and droop.

Delphinium carolinianum
Ranunculaceae

prairie larkspur

HABITAT Dry prairies, savannas, woodlands

BLOOM Summer

DESCRIPTION Slender, erect perennial, unbranched, short-hairy, to 3½ ft. tall

FLOWER White to pale blue, irregular, collection of five petal-like sepals, the upper one with a long tubular spur in back that curves upward, and four true petals that are comparatively small, the lower ones hairy on the inner surface

LEAF Basal and some alternate on stem, finely divided, palmately lobed, lobes narrow, to 4 in. long and wide

FRUIT Three-chambered, cylindrical capsule with finely pointed tips, to ⅔ in. long

Delphinium is from the Latin *delphinus* for "dolphin," an allusion to the shape of its flower bud. Its common name references the flower's spur, suggestive of the long hind claw on the feet of larks. There are several subspecies; the one in Minnesota is subsp. *virescens*. Some consider it a distinct species.

Hepatica americana
Ranunculaceae

round-lobed hepatica

HABITAT Dry to moist forests

BLOOM Spring

DESCRIPTION Erect perennial herb, unbranched, hairy (especially when young), to 6 in. tall

FLOWER White, but often purplish pink, to 1 in. wide, with 5–11 petal-like sepals in a ring around a center of numerous stamens, true petals absent, solitary on hairy stalks, several per plant

LEAF Basal, several per plant, three-lobed, each rounded at the tip, thick-textured, mottled on the surface, purplish underneath, persists through winter

FRUIT Flattened, elliptic nutlet

Some consider this species and sharp-lobed hepatica (*H. acutiloba*) (both also known as liverleaf) as varieties of *H. nobilis*. Each occur in the state. The main differences between the two species are the shape of the leaf lobes and habitat preference. Round-lobed hepatica tends to prefer drier sites. Both were used to treat ailments of the liver. *Hepatica* is from *hepaticus*, "of the liver."

Fragaria virginiana
Rosaceae

wild strawberry

HABITAT Prairies, savannas, open woodlands, forest edges, fields, roadsides

BLOOM Spring

DESCRIPTION Low-growing perennial herb producing runners that root at the tips to form colonies of new plants, to 6 in. tall

FLOWER White, to 1 in. wide, five-petals encircle a cluster of yellow stamens and pistils on a receptacle that will develop into a strawberry, several in a group at tip of flowering stem

LEAF Compound with three leaflets, mostly oval-shaped, sharply toothed, to 3 in. long and half as wide

FRUIT Fruiting receptacle (strawberry) fleshy, round, and red, with numerous dark, seedlike fruits each sunken in a pit

Fruits are small and sweet, much sweeter than garden strawberries. Another native, the woodland strawberry (*F. vesca*), is cone-shaped like the garden strawberry, with the "seeds" positioned on the surface (not in pits). The Ojibwe named it *ode'imîn* (heart berry), a reference to its shape.

Galium boreale
Rubiaceae

northern bedstraw

HABITAT Dry to wet sites, forests, various wetlands, thickets, ditches

BLOOM Summer

DESCRIPTION Erect perennial herb, branched with four-sided stem, usually smooth and hairless, occasionally hairy, to 3 ft. tall

FLOWER White, short, tubular, with four-pointed spreading lobes, each on stalks in clusters atop plant and upper leaf axils, to ¼ in. wide

LEAF Narrow, slightly widest near base, hairy along margins, to 2 in. long and ¼ in. wide, bearing three prominent veins, in groups of four leaves in evenly spaced whorls

FRUIT Ball-shaped, two-lobed (appears as two separate balls), hairy or not

Northern bedstraw is one of the showier bedstraws. It is in the madder plant family, as is coffee. Perhaps not surprisingly, the roasted seeds of a relative, cleavers (*G. aparine*), have been considered by some to be the best substitute for coffee of our northern plants.

Comandra umbellata
Santalaceae

bastard toadflax

HABITAT	Dry prairies, savannas, dunes, open woodlands, rocky shores
BLOOM	Spring, summer
DESCRIPTION	Erect, colonial perennial herb, hairless, partially parasitic, to 18 in. tall
FLOWER	White, bell-shaped, with five petal-like sepals, to ¼ in. long
LEAF	Elliptic, hairless, untoothed, to 2 in. long and ½ in. wide, alternate
FRUIT	Single-seeded with a thin, fleshy outer coat, reported to be somewhat edible

Although producing some of its food through photosynthesis, bastard toadflax also parasitizes several different plant species through a connection with their roots. It is one of the few (only?) plant species that occurs coast to coast, linked by adjoining counties across the entire United States. Interestingly, it also occurs in the Balkan Peninsula. It is an alternate host for a blister rust that affects pine trees.

flower

Mitella diphylla
Saxifragaceae

bishop's cap

HABITAT Moist forests, ravines, streambanks

BLOOM Spring

DESCRIPTION Erect perennial herb, unbranched, hairy, flower stalk to 1 ft. tall

FLOWER White, with five deeply fringed petals, to ¼ in. wide, several on slender upright stalk

LEAF Basal, with a single pair on the flowering stem; basal leaves with up to five sharp lobes, to 4 in. long and about as wide, stem leaves three-lobed, opposite at midstem, to 2½ in. long

FRUIT Capsules, splitting to expose a collection of shiny black seeds

Flowers resemble snowflakes, but one of the species' common names (miterwort), as well as its scientific one, refers to its fruit shape, which is reminiscent of a miter, a cap worn by bishops. The seeds are spread when rain splashes them out of the cap. A close relative, naked miterwort (*M. nuda*), has greenish flowers and a leafless flowering stalk. It is mostly in the northern counties.

Trillium cernuum
Trilliaceae

nodding trillium

HABITAT Moist forests, stream terraces, ravines, slopes

BLOOM Spring, summer

DESCRIPTION Erect perennial herb, unbranched, hairless, to 20 in. tall

FLOWER White, with three triangular petals upcurved at tips, on a long stalk that bends downward, often positioning the single flower below the leaves, to 2 in. wide

LEAF Technically leaflike bracts, broad, sharp-tipped, hairless, to 5 in. long and about as wide, in a whorl of three atop the stem

FRUIT Dark red, fleshy, berrylike seeds, dispersed by ants

Nodding trillium is not the showiest or largest trillium in the state but is clearly the most widespread. Another common name for the plant, birthwort, has been attributed to various Native Americans using it to facilitate childbirth. Trilliums have some of the largest chromosomes of all living things, six times as long as the human chromosome.

Pink to Red Flowers

Allium stellatum
Alliaceae

prairie onion

HABITAT Dry to moist prairies, savannas, fields, rock outcrops

BLOOM Summer

DESCRIPTION Erect perennial herb, single-stemmed, hairless, to 15 in. tall

FLOWER Pinkish purple, about ¼ in. wide, bearing six tepals, numerous in a rounded head about 2 in. wide

LEAF Narrow, grasslike, from near base of plant, about as long as flowering stalk, hairless, often withered by flowering time, emits onion smell when bruised

FRUIT Rounded capsules with black seeds

A somewhat similar onion in the southern half of the state, wild garlic (*A. canadense*), often has up to five flowers and numerous bulblets in its flowering head. Prairie onion is unusual in that it has two separate concentrations of occurrence within its overall range. One centers in the prairies of Minnesota and the other is farther south in calcareous glades of the Ozarks.

Asclepias incarnata
Apocynaceae

swamp milkweed

HABITAT Marshes, wet meadows, shorelines, ditches

BLOOM Summer

DESCRIPTION Erect perennial herb with milky sap, hairless, to 4 ft. tall

FLOWER Pink, about ¼ in. wide, consisting of five reflexed petals, pinkish white hoods, and incurved horns, numerous in somewhat flat-topped clusters to 3 in. wide

LEAF Narrowly lance-shaped, entire, hairless, to 6 in. long and 1 in. wide, opposite

FRUIT Narrow upright pods with smooth surfaces, to 4 in. long, seeds within have silky tufts of hairs that assist in wind dispersal

Of the several species of milkweed found in Minnesota, swamp milkweed is the one most likely to be found in wet sites. It is also one of the most common species. The Latin genus name is from Asclepius, Greek god of medicine and healing. The epithet *incarnata* references the flesh color of the flowers.

Asclepias syriaca
Apocynaceae

common milkweed

HABITAT Prairies, fields, roadsides, savannas, ditch banks

BLOOM Summer

DESCRIPTION Upright perennial herb, unbranched, short-hairy, with milky sap and long rhizomes, to 6 ft. tall

FLOWER Pale pink, consisting of five reflexed petals, pinkish white hoods, and incurved horns, to ½ in. wide, numerous in somewhat rounded clusters to 4 in. wide, fragrant

LEAF Broad, elliptic, densely hairy below, to 1 ft. long and 4 in. wide, in opposite pairs on stem

FRUIT Spreading pods to 4 in. long, surface hairy with soft prickles, seeds within have silky tufts of hairs that assist in wind dispersal

This is the most common milkweed in the state. It provides necessary food for monarch caterpillars and is critical to the species' survival. The fluffy seeds were once used as a filler for life jackets. Prairie milkweed (*A. sullivantii*) is similar but is hairless and mostly restricted to the southern counties.

Cirsium arvense
Asteraceae

Canada thistle, creeping thistle

HABITAT Fields, roadsides, ditches, fencerows, rights-of-way

BLOOM Summer

DESCRIPTION Erect perennial herb, colonial, to 4 ft. tall

FLOWER Pinkish purple, disk flowers only, in relatively small, crowded heads, each about ¾ in. wide, male and female flowers mostly separate on different plants

LEAF Lobed, mostly with sharp spines at tips, hairiness variable, alternate, to 6 in. long and 2 in. wide

FRUIT Seedlike, each bearing a long tuft of hairs at one end

Canada thistle is not native to Canada or North America. It's of Eurasian origin but has spread around the world as one of the most noxious weeds. It is certainly one of Minnesota's most invasive thistles. It forms large colonies by spreading roots as well as by being a prolific seed producer. Eradicate where possible.

Cirsium muticum
Asteraceae

swamp thistle

HABITAT Seeps, fens, swamps, wet meadows

BLOOM Summer, fall

DESCRIPTION Tall biennial herb, stem ridged, mostly spineless with sparse cobwebby hairs, to 8 ft. tall

FLOWER Pinkish purple, disk flowers only, numerous in widely spreading heads to 1½ in. wide, the little bracts on head below flowers mostly spineless, with cobwebby hairs and often sticky to the touch

LEAF Deeply lobed, with relatively soft spines, alternate, to about 1 ft. long and 4 in. wide

FRUIT Seedlike, each bearing a long tuft of hairs at one end

There are several species of thistles (*Cirsium*) in Minnesota. Swamp thistle is native and lacks the invasive nature of most of the nonnative ones. It is an important nectar source for pollinators, and its seeds are valued by birds, especially finches. It's apparently absent from the southwest corner of the state.

Echinacea angustifolia
Asteraceae

narrow-leaved purple coneflower

HABITAT Dry prairies, barrens

BLOOM Summer

DESCRIPTION Erect perennial herb, unbranched, rough-hairy, to 2 ft. tall

FLOWER Pinkish purple, in a single composite head per stem, to 2½ in. wide, outer ring of up to 20 ray flowers spreading or drooping, each about 1½ in. long, central disk cone-shaped, orangish red, with numerous tiny and tubular flowers

LEAF Mostly basal and alternate on lower stem, narrow, rough-hairy, three-veined, to 6 in. long, smaller up the stem

FRUIT Small, dry, one-seeded nutlet with angled sides

Mainly a species of the Great Plains, this coneflower is found primarily in the western half of the state. One source about historic uses of plants by Native Americans says that this species "seems to have been used as a remedy for more ailments than any other plant." In recent times, echinacea is marketed for the treatment of colds.

Eutrochium maculatum
Asteraceae

spotted Joe-Pye weed

HABITAT Marshes, bogs, fens, seep springs, wet prairies, ditches

BLOOM Summer

DESCRIPTION Erect, unbranched (below flower branches) perennial herb, hairy or nearly hairless, to 8 ft. tall, colonial

FLOWER Pinkish purple, disk flowers only, tubular, five-lobed, to 20 per compact head, the heads about ¼ in. wide, in a somewhat flat-topped branched cluster atop the main stem, no ray petals

LEAF Lance-shaped, narrowed at both ends, sharply toothed, veins deeply impressed, to 9 in. long, grouped in whorls of three to six on a vertical stem that's commonly purple-spotted

FRUIT Slender, one-seeded, with tuft of long hairs at one end that helps it float in the wind

Many stories exist about the origin of the name Joe-Pye weed. It was most likely named after Joseph Shauquethqueat, or "Joe Pye," a Mohican leader from the late 18th century. He lived in western Massachusetts, where the name of the plant was likely first used.

rough blazing star

Liatris pycnostachya
Asteraceae

prairie blazing star

HABITAT Moist prairies, meadows, fields

BLOOM Summer

DESCRIPTION Erect perennial herb, unbranched, variously hairy, to 5 ft. tall

FLOWER Pink, disk flowers only, up to eight small, tubular five-lobed flowers in heads to ½ in. wide, the tiny hairy bracts on head below flowers with pointed tips that curve outward, densely arranged on a single spike, blooming from the top down

LEAF Narrow, grasslike, alternate, the lowest ones to 1 ft. long and ½ in. wide, becoming progressively smaller up the stem

FRUIT Small, dry, wedge-shaped nutlet with silky hairs that facilitate wind dispersal

One of several *Liatris* species native to Minnesota. Given its densely packed, long spike of small flower heads, it is one of the easiest of the group to identify in the field. Flowers of the 1 in. wide heads of the rough blazing star (*L. aspera*, see inset) number as many as 40. Both are popular in the nursery trade.

Brasenia schreberi
Cabombaceae

watershield

HABITAT Acidic water in lakes, ponds, swamps, slow-moving streams

BLOOM Summer

DESCRIPTION Aquatic, colonial perennial rooted in soil of still bodies of water

FLOWER Pinkish red to purple, about ¾ in. wide, with three recurved petals and three similar sepals, on a stalk held just above the water during the day, withdraws under the surface each evening

LEAF Floating on water surface, oval, somewhat football-shaped, to 5 in. long and 3 in. wide, undersurface purple, and has a slimy, jellylike coating, as does the stalk

FRUIT Club-shaped, one- or two-seeded, in clusters

The mucilaginous coating that covers most of the plant helps protect it from feeding insects. In some Asian countries it is eaten as a vegetable. During the first day the flower opens, it is functionally male; on the second day, it is functionally female. In some populations, this order may be reversed.

Trifolium pratense
Fabaceae

red clover

HABITAT Pastures, fields, roadsides, ditches, disturbed areas

BLOOM Summer, fall

DESCRIPTION Erect or ascending perennial, branched, hairy, to 2 ft. tall

FLOWER Pinkish red to purplish, to 1½ in. wide, tubular, five-petaled, with typical pea flower shape, several in a rounded head positioned immediately above a pair of small leaves

LEAF Compound, alternate, with three broadly elliptic leaflets, unlobed, margins smooth or finely toothed, often with a light green chevron patch, to 2½ in. long and 1 in. wide

FRUIT Small pod hidden within the withered flower

This introduced Eurasian clover is a forage crop for livestock. Because it also fixes nitrogen, it is often planted as a cover crop for soil health. Regarding health, one study on female lab rats showed that red clover isoflavones may help reduce bone loss. Several other *Trifolium* clovers in Minnesota are nonnative and weedy.

Monarda fistulosa
Lamiaceae

wild bergamot

HABITAT Prairies, savannas, fields, open woodlands, thickets, roadsides, usually in rather dry conditions

BLOOM Summer

DESCRIPTION Erect perennial herb, aromatic, branched, finely hairy, with square stems to 4 ft. tall

FLOWER Pinkish, tubular, two-lipped, the upper one narrow and unlobed, hairy, the lower one three-lobed, to 1½ in. long, several ringing a dome-like head at stem tips

LEAF Lance-shaped with toothed margins, on short stalks, opposite, usually hairy, to 5 in. long and 2 in. wide

FRUIT Dry, with usually four nutlets

Wild bergamot has a purported similarity of scent with a type of sour orange grown near the town of Bergamo, Italy. Also called bee balm for being attractive to bees. It is very minty and thus used in potpourri and sachets. Historically it was used to treat a variety of medical conditions. The Ojibwe have used it to cure catarrh and bronchial issues.

germander

Stachys pilosa
Lamiaceae

marsh hedgenettle

HABITAT Marshes, wet prairies, meadows, stream borders, shores, ditches

BLOOM Summer

DESCRIPTION Erect perennial herb, unbranched, square-stemmed, with spreading hairs on angles as well as sides, to 3 ft. tall

FLOWER Pink with dark purple spots, to ½ in. long, tubular, two-lipped, upper lip hoodlike, lower three-lobed with the lower one the longest, in several whorls on upper stem

LEAF Lance-shaped, deeply veined, margins with short, sharp teeth, opposite, lacks pleasant minty aroma, to 4 in. long and about 2 in. wide

FRUIT Dry, capsule-like, separating into four nutlets

Sometimes known as *S. palustris*, but technically that species is European and not known in the state. *Stachys* is Greek for "spike," probably in reference to the plant's arrangement of flowers. Interestingly, a person named Stachys is mentioned in Romans 16 of the Bible. The flower of germander (*Teucrium canadense*, see inset) is similar but the upper hood is replaced by two pointed lobes.

Mirabilis nyctaginea
Nyctaginaceae

wild four-o'clock

HABITAT Moist and dry prairies, roadsides, fields, railroads, disturbed sites

BLOOM Summer

DESCRIPTION Erect perennial herb, branched, stem predominantly hairless and angled, to 3 ft. tall

FLOWER Pink, to ½ in. wide, petals absent, but with five petal-like sepals notched at the tips, in clusters at stem tips

LEAF Triangular with a somewhat heart-shaped base, opposite, to 4 in. long and 3 in. wide

FRUIT Nutlets, hairy and ribbed

So named for the opening of its flowers in late afternoon, wild four-o'clock is related to the garden version (*M. jalapa*) from South America. Wild four-o'clock can be invasive, colonizing sites far from its natural range in the Great Plains. Caterpillars of the micromoth (*Neoheliodines nyctaginella*) feed only on wild four-o'clocks and related species.

Cypripedium acaule
Orchidaceae

pink lady's slipper

HABITAT Dry or wet acidic forests, bogs, swamps, sphagnum moss hummocks

BLOOM Spring

DESCRIPTION Erect perennial herb, unbranched, hairy, to 18 in. tall

FLOWER Pink, three-petaled, one shaped like a pouch with a center cleft and reddish veins, to 2½ in. long, the other two purplish, lance-shaped with a slight twist, to 1½ in. long, solitary on a single stalk coming from the center between the basal leaves

LEAF Basal, in a pair, elliptic, unlobed, hairy, ribbed, to 8 in. long and 5 in. wide

FRUIT Ascending capsule with thousands of tiny, dustlike seeds

There are five species of lady's slippers in the state, and this is the most common, especially northward. The genus name is from the Greek *Kypris*, from Cypris, the goddess of love and beauty, and the Latin *pedis* for "foot." It is also known as moccasin flower for the shape of the large slipper-shaped lower petal.

Cypripedium reginae
Orchidaceae

showy lady's slipper

HABITAT Fens, seepage swamps, sedge meadows

BLOOM Summer

DESCRIPTION Erect, unbranched perennial, hairy, to 3 ft. tall, often growing in clumps

FLOWER Pink and white, one to three flowers atop stem, three-petaled, the pinkish pouch to 2½ in. long, the other two lateral petals white, untwisted, to 2½ in. long, three sepals, white, appearing as two (two are fused)

LEAF Alternate, oval, noticeably ribbed, covered with short glandular hairs, to 10 in. long and 5 in. wide

FRUIT Erect capsule with thousands of tiny, dustlike seeds

This is the state's largest native orchid and the official state flower, making Minnesota the only state to have an orchid so designated. Some people have sought to transplant them from the wild into their home gardens. This practice negatively impacts wild populations. It is best to leave them in place for all to enjoy.

Phlox pilosa
Polemoniaceae

prairie phlox

HABITAT Moist and dry prairies, savannas, barrens, open woodlands

BLOOM Spring, summer

DESCRIPTION Erect perennial herb, multi-stemmed, mostly unbranched except above, hairy, to 2 ft. tall

FLOWER Pink, to ¾ in. wide, with five flaring lobes at tip of ⅔ in. long tube, the base of each often with darker pink markings, in a rounded cluster atop the plant

LEAF Narrowly lance-shaped, unlobed, toothless, usually hairy, opposite, pairs positioned at 90 degrees to one another

FRUIT Rounded capsule with seeds thrown some distance from plant upon drying

Phlox is Greek for "flame," perhaps referring to the plant's twisted flower bud, shaped like a tiny torch. As natural prairie disappears in the state, so does this plant. In turn, animals that depend on it are also affected, such as the rare phlox moth (*Schinia indiana*), whose caterpillars feed only on flowers of prairie phlox.

Persicaria amphibia
Polygonaceae

water smartweed

HABITAT Lakes, ponds, slow-moving streams and rivers, marshes, ditches

BLOOM Summer, fall

DESCRIPTION Erect to sprawling aquatic perennial herb, branched or not, hairy or hairless, to 4 ft. tall

FLOWER Pink, five tepals, to ⅛ in. wide and ¼ in. long, packed into either egg-shaped clusters (var. *stipulacea*, pictured, has floating leaves) or longer cylindrical clusters (var. *emersa* grows on exposed soil)

LEAF Floating on water's surface, or not, lance-shaped or elliptic, alternate and with a papery sheath where the leaf stalk attaches to a knotlike swelling on the stem, hairy or not

FRUIT Dry nutlet, two- or three-angled, often hidden in dried flowers

When eaten, some species of smartweeds have a peppery taste that "smarts" the tongue. Also known as knotweeds because of the swollen knots on the stem where the leaves attach. Other family members include buckwheat and rhubarb. Seeds of water smartweed are desirable food for wildlife, particularly waterfowl.

Aquilegia canadensis
Ranunculaceae

wild columbine

HABITAT Rock outcrops, steep wooded slopes, streambanks, terraces

BLOOM Spring, summer

DESCRIPTION Multi-branched perennial, stems hairless, to 30 in. tall, basal leaves plus alternately attached stem leaves

FLOWER Red and yellow, about 2 in. long, shaped unlike anything else in state, nods from arching stem tips, bears long vertical spurs with club-shaped tips, five yellow petals, five red petal-like sepals, numerous yellow stamens exerted from throat

LEAF Doubly compound, three leaflets, each with three deeply lobed subleaflets about 2 in. long and wide

FRUIT Pods have a long tail (beak), within which are shiny black seeds

Wild columbine is the only naturally occurring columbine in eastern North America. It likely occurs in every county of the state. The Latin *columbinus* means "dove-like," and the name reflects the allusion of doves portrayed in the inverted flower.

Rosa blanda
Rosaceae

smooth rose

HABITAT Open woodlands, prairies, savannas, bluffs, barrens, thickets, fields, roadsides

BLOOM Summer

DESCRIPTION Erect woody shrub, branched or not, prickles on older wood at base, usually smooth on upper stems, to 4 ft. tall

FLOWER Pink, with five rounded petals and numerous yellow stamens, to 3 in. wide, occurring singly or few on stem tips

LEAF Compound, alternately attached to stem, usually with seven leaflets, the latter elliptic, coarsely toothed, to 1½ in. long and about 1 in. wide

FRUIT Rounded rose hips, red when ripe

Roses are well known around the world, and Minnesota has four native species. Although called smooth rose, as its upper stems are relatively thorn-free, the lower stems can be quite prickly. The similar prairie rose (*R. arkansana*) is prickly throughout. The Ojibwe have used rose hips for relieving heartburn and as an eye medicine.

Orange Flowers

Asclepias tuberosa
Apocynaceae

butterfly milkweed

HABITAT Prairies, old fields, savannas, roadsides, especially sandy sites

BLOOM Summer, fall

DESCRIPTION Spreading, coarsely hairy perennial herb to 3 ft. tall, with clear sap

FLOWER Orange, red, or yellow, about ½ in. wide, consisting of five reflexed petals, hoods, and incurved horns, numerous in somewhat rounded clusters to 3 in. wide

LEAF Narrow, lance-shaped, hairy, mostly alternate on stem, to 4 in. long and 1 in. wide

FRUIT Upright, lance-shaped pods, finely hairy, to 6 in. long; seeds within have silky tufts of hairs that assist in wind dispersal

The brilliantly colored flowers of butterfly milkweed have attracted considerable interest for use as a landscaping plant, especially in native-plant gardens. In the wild, occurrence in the state is mostly in the southeastern counties. The plant lacks milky sap, an unusual trait for a milkweed. Like other milkweeds, it is a food host for monarch caterpillars.

yellow touch-me-not

Impatiens capensis
Balsaminaceae

jewelweed

HABITAT Seeps, swamps

BLOOM Summer, fall

DESCRIPTION Branching, erect annual herb, hollow-stemmed, hairless, to 4 ft. tall

FLOWER Orange, cone-shaped like a cornucopia, to 1 in. long, five-lobed, open in front and narrowing to a small spur in back, spotted, dangles from upper leaf axils

LEAF Alternate, elliptic, with wavy or bluntly toothed margins, to 5 in. long and 3 in. wide

FRUIT Green cylindrical capsule

A leaf submerged in water shows why this plant is named jewelweed, as it takes on a silvery sheen. Also known as touch-me-not because of the explosive nature of the ripe seed capsule when disturbed. *Impatiens* is the Latin word for "impatient," alluding to the seed, which "can't wait" to leave the capsule. The similar yellow touch-me-not (*I. pallida*, see inset) grows in southeastern Minnesota.

Michigan lily

Lilium philadelphicum
Liliaceae

wood lily

HABITAT Dry prairies, meadows, fields, open woodlands

BLOOM Summer

DESCRIPTION Erect perennial herb, unbranched, hairless, to 3 ft. tall

FLOWER Orange, to 3 in. wide, consisting of six spoon-shaped and petal-like tepals with purplish brown spots at their base, facing upward, one to three atop the stalk

LEAF Narrow, typically in whorls of 3–11 or alternate on lower stem, to 4 in. long and 1 in. wide

FRUIT Cylindrical capsule to 2 in. long

Wood lily, also known as prairie lily, is truly a gem of a wildflower and is widely admired, even considered the provincial flower of Saskatchewan. The other native *Lilium* species in Minnesota, Michigan lily (*L. michiganense*, see inset), is a taller plant with multiple nodding flowers per stem. It occurs primarily in the eastern half of the state.

Castilleja coccinea
Orobanchaceae

Indian paintbrush

HABITAT Moist prairies (especially sandy), savannas, meadows, sandy and gravely shorelines

BLOOM Spring, summer

DESCRIPTION Erect biennial herb, partially parasitic, hairy, to 2 ft. tall

FLOWER Greenish yellow, tubular, to ¾ in. long, mostly obscured by large, variously lobed red, orange, or yellow bracts

LEAF Three-lobed, stalkless, alternate, hairy, to 3 in. long

FRUIT Smooth, oblong capsule

The modified leaves (bracts) below and surrounding the flowers are the colorful parts. Widespread in the state except missing in the southwestern counties, it is one of only a few species of paintbrush in the eastern United States (many occur in the West). Indian paintbrush is partially parasitic, obtaining some but not all of its nutrition from other plants through a connection with their roots.

Yellow Flowers

Acorus americanus
Acoraceae

American sweetflag

HABITAT Marshes, pond and lake margins, seep springs, wet prairies

BLOOM Spring, summer

DESCRIPTION Erect, unbranched, hairless perennial, to 6 ft. tall

FLOWER Yellowish green, very tiny, hundreds tightly packed on a fingerlike spike, to 3 in. long and ¾ in. wide, arising from the lower half of leaf

LEAF Slender, bright green, and sword-shaped, like an iris or a cattail, to 4 ft. tall and ¾ in. wide, with multiple veins parallel to the larger midvein

FRUIT One- to six-seeded small berry

When bruised, the foliage of sweetflag is pleasantly aromatic. It was once highly prized for its medicinal properties. A similar-looking sweetflag from Europe (*A. calamus*) has been introduced into North America but is apparently absent or quite rare in Minnesota. The native one has fertile flowers while the introduced one does not.

Zizia aurea
Apiaceae

golden alexanders

HABITAT Moist prairies, sedge meadows, dry and moist woodlands, thickets, creekbanks

BLOOM Spring, summer

DESCRIPTION Erect perennial herb, somewhat-branched, hairless, colonial, to 3 ft. tall

FLOWER Yellow, five-petaled, to ⅛ in. wide, all on short stalks except for a central stalkless one, up to 25 in somewhat flat-topped, small clusters that together make up a larger flattish cluster to 3 in. wide

LEAF Compound, basal leaves twice three-parted, stem leaves sometimes less parted, leaflets lance-shaped, sharply toothed, overall outline to 6 in. long and about as wide

FRUIT Oval, flattened, ribbed, with two seeds

Caterpillars of the black swallowtail (*Papilio polyxenes*) feed on foliage of this species, and golden alexanders mining bees (*Andrena ziziae*) feed almost exclusively on its pollen. Most or all basal leaves in the state's other species of *Zizia*, heart-leaved golden alexanders (*Z. aptera*), are heart-shaped.

Bidens cernua
Asteraceae

nodding bur-marigold

HABITAT Mudflats, shorelines, swamps, fens, ditches

BLOOM Summer, fall

DESCRIPTION Erect, branching, mostly hairless annual herb, to 3 ft. tall

FLOWER Yellow, in composite heads to 2 in. wide, outer ring of usually eight ray flowers, or absent, flowers of center disk numerous, tiny, and tubular

LEAF Opposite, hairless, coarsely toothed, with bases fused to the stem, to 5 in. long and 1 in. wide

FRUIT Flattened and brown with four barbed awns capable of sticking to fur and clothing (and skin!)

Part of a group known as beggar-ticks, alluding to the fruits sticking to clothing and fur, as if "begging" a ride. The Latin name *cernua* means "nodding," a reference to the flowering heads that nod with age. Some plants bloom when only 1 or 2 in. tall.

Helenium autumnale
Asteraceae

common sneezeweed

HABITAT Wet meadows, fields, fens, swamps, shorelines

BLOOM Summer, fall

DESCRIPTION Erect perennial herb, branched, variously hairy, with thin blades (wings) of tissue extending down the stem from leaf bases, to 5 ft. tall

FLOWER Yellow, in composite heads arranged at branch tips, to 2 in. wide, outer ring of up to 20 ray flowers with notched tips, spreading or drooping, each to ¾ in. long; central disk rounded, yellow, with hundreds of tiny, tubular flowers

LEAF Lance-shaped, alternate, tapered at both ends, with small, widely scattered teeth on the margins, to 5 in. long and ¾ in. wide

FRUIT Narrow, wedge-shaped, ribbed, about 1⁄16 in. long

Sneezeweed can cause sneezing, but not from its pollen, which is too heavy to be windborne and thus would not be inhaled. Its common name reflects the traditional use of its powdered leaves to make snuff, which brings on sneezing, and the health benefits some believe can be derived from it.

Helianthus giganteus
Asteraceae

giant sunflower

HABITAT Marshes, wet prairies, fens, swamps, wet forests

BLOOM Summer, fall

DESCRIPTION Erect perennial herb, stem purple or green, usually with spreading hairs, to 10 ft. tall

FLOWER Yellow, in composite heads arranged on tips of spreading branches, to 3 in. wide, outer ring of up to 20 showy, sterile ray flowers, each to about 1¼ in. long, central disk yellow with dozens of tiny, tubular flowers

LEAF Lance-shaped, mostly alternate on upper stem, opposite on lower, tapered at both ends, usually with coarse teeth on the margins, surface rough, to 6 in. long and 1½ in. wide

FRUIT Small, dry, one-seeded nutlet, smaller than that of garden sunflower

Helianthus means "sun flower," from the combination of the Greek *helios* (sun) and *anthos* (flower). There are several species in the state. Smooth oxeye (*Heliopsis helianthoides*) is a common sunflower-like plant, but each of its outer ray flowers is fertile, producing a seed (sterile in sunflower).

Matricaria discoidea
Asteraceae

pineapple-weed

HABITAT Disturbed sites, gravel roads, cracks in pavement, parking lots, compacted soils

BLOOM Spring to fall

DESCRIPTION Erect annual herb, branched, mostly hairless, to 1 ft. tall (but usually much shorter)

FLOWER Yellowish green, tubular, abundant in cone-shaped heads, ray flowers absent, to ⅓ in. wide, on upright stalks from upper leaf axils

LEAF Feathery, deeply divided in several linear segments, alternate, to 2 in. long and ¾ in. wide

FRUIT Tiny, dry, cylindrical nutlets

Introduced from the western states, it mostly avoids natural areas, usually confining itself instead to highly disturbed sites. The plant has a pleasant pineapple scent detected either by walking on it or crushing it with your fingers. Various Indigenous peoples have used the plant for treating gastrointestinal maladies or as a perfume or an insecticide.

Packera paupercula
Asteraceae

balsam ragwort

HABITAT Moist prairies, meadows, fens, shores

BLOOM Spring, summer

DESCRIPTION Erect perennial herb, unbranched, mostly hairless, but may have cobwebby hairs in leaf axils and lower stem, to 1½ ft. tall

FLOWER In compact heads to ¾ in. wide, composed of an outer ring of up to 13 yellow ray flowers and a central disk of several tiny yellow and tubular flowers, arranged in spreading clusters atop the stem

LEAF Mostly basal, oval to lance-shaped with small, rounded to sharp teeth along margins, to 3 in. long and ¾ in. wide, stem leaves small, scattered, and deeply lobed

FRUIT Tiny, dry, with tuft of long, silky hairs at one end

There are several species of ragwort in Minnesota, including the similar prairie ragwort (*P. plattensis*). It is usually quite hairy and grows in dry prairie habitats. The common name appears to refer to the plant's leaves, especially the "ragged" stem leaves. The term *wort* means "plant," from the Old English *wyrt*.

Rudbeckia hirta
Asteraceae

black-eyed Susan

HABITAT Prairies, savannas, fields, meadows, roadsides

BLOOM Summer, fall

DESCRIPTION Erect annual or biennial herb, mostly unbranched except possibly for upper plant, coarsely hairy, to 2½ ft. tall

FLOWER In compact heads to 3 in. wide, composed of an outer ring of up to 20 yellow ray flowers and a central cone-shaped disk of several tiny, blackish brown tubular flowers, the heads atop plant or a few from stalks in leaf axils

LEAF Basal leaves stalked, stem leaves usually not, lance-shaped, toothless or with few teeth, hairy, alternate, to 7 in. long and 2 in. wide

FRUIT Narrow, dark, wedge-shaped, four-angled nutlet

This wildflower is also a popular garden plant. The common name refers to the dark "eye" of the flowering head and may originate from an early 18th century English poem titled "Sweet William's Farewell to Black-Ey'd Susan." Perhaps the dark disk of the flowering head reminded the colonists of the poem's lady.

Solidago altissima
Asteraceae

tall goldenrod

HABITAT Prairies, savannas, fields, roadsides, meadows, thickets

BLOOM Summer, fall

DESCRIPTION Erect perennial herb, unbranched except for flowering branches, hairy, rhizomatous, to 6 ft. tall

FLOWER Yellow, to ⅓ in. wide, comprising an outer ring of up to 15 ray flowers and a central disk of three to eight tiny, tubular flowers, in a spreading, pyramidal arrangement atop the stem

LEAF Elliptic to lance-shaped with a few small teeth, three-veined, upper surface commonly rough, alternate, to 6 in. long and 1 in. wide

FRUIT Brown, narrowly wedge-shaped nutlet (seed within)

One of the more common of the several species of goldenrods in the state. They get a bad rap for causing hay fever, but their pollen is too heavy to be airborne and inhaled. The most likely cause of hay fever is ragweed (*Ambrosia* species). Its pollen is dustlike and wind-dispersed at the time goldenrods are in bloom.

Taraxacum officinale
Asteraceae

common dandelion

HABITAT Lawns, fields, pastures, roadsides, waste areas

BLOOM Mostly spring, through fall

DESCRIPTION Erect perennial herb, unbranched, hairless or sparsely so, milky sap, to 15 in. tall

FLOWER Yellow, to 2 in. wide, usually with up to 100 ray flowers, disk flowers absent, crowded in a composite head, solitary at tip of a vertical hollow stem

LEAF All basal, elliptic or widest toward tip, with deeply triangular lobes and irregular teeth, to 10 in. long and 3 in. wide

FRUIT Small, one-seeded, brownish nutlet attached to a plume of bristles, together with others form a globe-shaped cluster (puffball) to about 1 in. wide

Dandelions were introduced from Eurasia but occur on all continents except Antarctica. The word *dandelion* appears to be from the French *dent de lion* (tooth of the lion), a reference to its toothed leaves, but in France today it's called *pissenlit* (and historically *pissabed* in England), names referring to the diuretic effects from eating the plant, leading to increased urination.

Tragopogon dubius
Asteraceae

goat's beard

HABITAT Fields, roadsides, pastures, waste areas, disturbed areas

BLOOM Spring, summer

DESCRIPTION Erect biennial herb, gray-green, cobwebby hairs when young, hairless at maturity, milky sap, to 3 ft. tall

FLOWER Pale yellow, to 2 in. wide, with 100 or more ray flowers, disk flowers absent, small leafy bracts of head longer than flowers, head solitary at tip of a vertical hollow stem, swollen at tip

LEAF Mostly basal and alternate, linear, grasslike, unlobed, mostly hairless, clasping at base, to 12 in. long and ¾ in. wide

FRUIT One-seeded, brownish nutlet attached to a plume of bristles, clusters in a large globe shape resembling a giant dandelion seed head, to 3 in. wide

This European native was named for its grayish, fluffy seed head, from the Greek *tragos* (goat) and *pogon* (beard). Flowers usually open in early morning and close around noon on sunny days, inspiring another common name, John-go-to-bed-at-noon.

berrylike seeds

Caulophyllum thalictroides
Berberidaceae

blue cohosh

HABITAT	Moist, rich woods, ravines, shaded slopes
BLOOM	Spring
DESCRIPTION	Erect perennial herb, hairless, to 2 ft. tall, stem green or purplish, often with whitish coating that rubs off
FLOWER	Yellow or greenish yellow, some with purple tinge, to ½ in. wide, six petal-like sepals (true petals obscure)
LEAF	Compound, two per plant if flowering or fruiting, one if not, leaflets several, to 3 in. long with pointed lobes
FRUIT	Fleshy, berrylike seeds, green at first, then dark blue when ripe

The Latin name *thalictroides* means "like *Thalictrum*," the genus of meadow rue. Leaves of these two unrelated plants have a similar appearance. The berrylike seeds may entice you to eat them, but they are reported to be poisonous. The Ojibwe have used the root for menstrual cramps.

Lithospermum canescens
Boraginaceae

hoary puccoon

HABITAT	Dry to moist prairies, savannas, open woodlands
BLOOM	Spring, summer
DESCRIPTION	Erect perennial herb, soft-hairy, in clumps, to 1½ ft. tall
FLOWER	Yellow, tubular, with five spreading lobes, to ½ in. wide, several in a flattish cluster atop the plant
LEAF	Narrow, unlobed, soft-hairy with prominent midvein, alternate, to 2 in. long and ½ in. wide
FRUIT	Hard, white nutlet

The genus name refers to its hard nutlets, from the Greek *lithos* (stone) and *sperma* (seed), and the specific epithet *canescens* (hoary) refers to its soft, grayish white foliage. The similar hairy puccoon (*L. caroliniense*) is rough-hairy. *Puccoon* is a common name used by some Indigenous people for plants used for dyes. Roots of hoary puccoon have been used to make a reddish dye.

Barbarea vulgaris
Brassicaceae

yellow rocket

HABITAT Roadsides, fields, riverbanks, waste areas, thickets

BLOOM Spring

DESCRIPTION Erect biennial herb, branched above, mostly hairless, stem somewhat angled, to 2½ ft. tall

FLOWER Yellow, four-petaled, to ⅓ in. wide, in compact clusters at branch tips

LEAF Basal and lower leaves deeply lobed and stalked, becoming progressively smaller, unlobed and stalkless upward, alternate, the larger to 8 in. long

FRUIT Narrow, upright pods splitting lengthwise into two halves

This early-blooming wildflower of the mustard family is introduced from Eurasia. The genus name honors Saint Barbara, patron saint of people involved with explosives, such as artillerymen and miners. The plant is said to have been useful in covering and healing wounds. The Latin epithet *vulgaris* means "common." Also called wintercress, as the basal leaves overwinter.

Clintonia borealis
Convallariaceae

bluebead lily

HABITAT Moist forests, swamps, bogs

BLOOM Spring

DESCRIPTION Erect perennial herb, unbranched, to 15 in. tall, colonial

FLOWER Yellow, bell-shaped, with six slender and flaring petal-like segments, to ¾ in. wide, in a cluster of up to six on a bare stem

LEAF Basal, glossy, and somewhat thick and fleshy, oblong, with fine hairs along the margins, to 8 in. long and 3 in. wide

FRUIT Round, deep blue berry, about ¼ in. wide

Contrary to its common name, bluebead lily is not a true lily. Its berries are an attractive porcelain blue and though they may look appetizing, they are said to be bitter-tasting and mildly toxic. The Ojibwe have used the roots to make tea to aid in childbirth.

Uvularia grandiflora
Convallariaceae

large-flowered bellwort

HABITAT Rich, moist forests, stream terraces, ravines, slopes

BLOOM Spring

DESCRIPTION Erect perennial herb, branched, arching, mostly hairless, often clump-forming, to 2 ft. tall

FLOWER Yellow, bell-shaped, to 2 in. long, six tepals, slightly twisted, drooping from stem tips

LEAF Elliptic, perfoliate (stem within blade tissue, as if pierced), unlobed, undersurface often hairy, alternate, to 4 in. long and half as wide

FRUIT Capsule, three-lobed with rounded edges

Uvularia is named after the uvula, that dangling tissue in the back of a person's throat. Its "pierced" leaves are distinctive. Leaves of the other bellwort in the state, sessile-leaved bellwort (*U. sessilifolia*), are not pierced. The root of large-flowered bellwort, known to the Ojibwe as *wabûckadji'bîk* (white root), has been used to treat stomach ailments.

Lotus corniculatus
Fabaceae

bird's-foot trefoil

HABITAT Roadsides, pastures, fields, disturbed sites

BLOOM Summer

DESCRIPTION Sprawling perennial herb, branched, variously hairy or hairless, angled, to 2 ft. tall

FLOWER Yellow, to ½ in. long, five-petaled with a typical pea family shape, the broad uppermost banner petal may be marked with red at the base, occurring in clusters on long stalks

LEAF Pinnately compound with five leaflets, the lowest two separated on the stalk from the crowded upper three, oval, to ¾ in. long and half as wide

FRUIT Narrow, pea pod–like, dark in color, to 1 in. long

Introduced from Eurasia as a cover crop and for forage, this invasive plant is a road-runner, often found along manicured highway verges. Bird's-foot trefoil is a member of the bean family and, despite its scientific name, is not closely related to American lotus (*Nelumbo lutea*), an aquatic plant with large, round leaves and whitish flowers.

Melilotus officinalis
Fabaceae

yellow sweet clover

HABITAT Fields, roadsides, waste areas, prairies; preferring dryish, calcareous soils

BLOOM Summer

DESCRIPTION Erect, usually biennial, branches ascending, hairless, to 5 ft. tall

FLOWER Yellow, five-petaled with typical pea family shape, to ¼ in. long, in loose clusters at tips of upright stalks

LEAF Compound, with three oval, finely tooted leaflets to 1½ in. long and ⅔ in. wide, alternate

FRUIT Strongly wrinkled single-seeded pod

Despite its common name, yellow sweet clover is not a species that most people recognize as clover. That name is usually reserved for the group that includes white and red clovers. Also known as melilot, it was introduced from Europe with good intentions but has become an invasive weed. White sweet clover (*M. albus*) is similar but has white flowers.

leaves
and bladders

Utricularia vulgaris
Lentibulariaceae

common bladderwort

HABITAT Lakes, ponds, swamps, slow-moving streams and rivers, ditches

BLOOM Summer

DESCRIPTION Floating aquatic perennial, carnivorous, stems to 3 ft. long

FLOWER Yellow, looks somewhat like a snapdragon, two-lipped, upper lip upright, the lower with a hump and a forward-projecting spur from beneath, to ¾ in. long, several on a stalk extending above the water's surface

LEAF Threadlike segments forking several times, bears small bladders (traps) to catch and digest tiny aquatic organisms

FRUIT Roundish capsule holds numerous tiny seeds

This carnivorous plant has tiny bladders with hinged trap doors that, when triggered, rapidly suck in unsuspecting organisms such as water fleas and mosquito larvae. The bladders contain secretions that break down prey so that nutrients can be absorbed. Some microorganisms, however, can live in the bladder unaffected.

starflower

Lysimachia thyrsiflora
Myrsinaceae

tufted loosestrife

HABITAT Swamps, bogs, fens, floodplains, lakeshores, ditches

BLOOM Summer

DESCRIPTION Erect perennial herb, unbranched, finely hairy on upper stem, to 2½ ft. tall

FLOWER Yellow, to ⅓ in. wide, with five to seven spreading narrow petals and a similar number of long-exerted yellow stamens, several in oval clusters on stalks from leaf axils at middle of plant

LEAF Slender, lance-shaped to linear, opposite, to 4 in. long and ½ in. wide, commonly dotted on the surfaces

FRUIT Round, dry capsules

The genus name is perhaps from *lysis* (a release from) and *mache* (strife), from the story of Lysimachus, king of Thrace, who pacified an angry bull that chased him by waving the plant before it. All the native loosestrife (*Lysimachia*) species in Minnesota have yellow or greenish flowers except starflower (*L. borealis* or *Trientalis borealis*, see inset), which has white flowers.

Nuphar variegata
Nymphaeaceae

yellow pond lily

HABITAT Ponds, lakes, calm water portions of rivers, often rooted in highly organic soil

BLOOM Summer

DESCRIPTION Aquatic perennial herb with thick, spongy rhizomes from which the leaves grow, the leaf stalks to 6 ft. long

FLOWER Yellow with maroon base, petals tiny and inconspicuous, usually six sepals showy, rounded, together forming what looks like a yellow cup or bowl, solitary on vertical stalks held well above the water level, to 2 in. wide

LEAF Shiny, oval to narrowly heart-shaped with cleft at base, usually floating level with surface (sometimes above during low water), the stalks flattened on one side, to 10 in. long and 6 in. wide

FRUIT Berrylike, somewhat fleshy, with numerous seeds, on straight stalks

This plant, along with water lily (*Nymphaea odorata*), are often called lily pads. The leaf stalks of yellow pond lily are attached to a thick, ropy looking rhizome. The Ojibwe call it *odîte'abûg* (flat heart leaf).

Oenothera biennis
Onagraceae

common evening primrose

HABITAT Prairies, old fields, fallow farm fields, roadsides, disturbed sites, usually on rather dry soils

BLOOM Summer, fall

DESCRIPTION Erect biennial herb, unbranched or branched above, smooth or hairy, to 6 ft. tall

FLOWER Yellow, to 2 in. wide, comprising four rounded petals with notched tips and four long, pointed, reflexed greenish sepals, borne on leafy spikes atop the stem

LEAF Lance-shaped, hairy, margins toothed or not, wavy, alternate, to 7 in. long and 2 in. wide, leaves all basal first year

FRUIT Narrow, cylindrical, dry capsule

This rather weedy native is by far the most common of the dozen or so evening primrose species in the state. Closed for most of the day, its flowers open quickly in the evening, sometimes in less than a minute. The colorful primrose moth (*Schinia florida*) hides inside the flowers during the day, and its caterpillars feed exclusively on them.

Cypripedium parviflorum
Orchidaceae

yellow lady's slipper

HABITAT Variety of forest sites in rather dry to wet soils, fens, swamps

BLOOM Spring, summer

DESCRIPTION Erect, unbranched perennial, hairy, to 2 ft. tall, often growing in clumps

FLOWER Yellow, three-petaled, one shaped like a pouch to 2½ in. long, the other two in lateral spiraled petals to 3 in. long

LEAF Alternate, lance-shaped to oval, noticeably ribbed, hairy, to 9 in. long and 5 in. wide

FRUIT Erect capsule with thousands of tiny, dustlike seeds

Minnesota has two varieties of yellow lady's slippers (var. *pubescens*, pictured) and (var. *makasin*). The latter's flower usually has a smaller pouch and darker-colored lateral petals. It is also found in wetter sites. The Ojibwe name is *ma'kasîn* (moccasin). The root is said to be a good remedy for "female troubles."

Pedicularis canadensis
Orobanchaceae

wood betony

HABITAT Dry to moist prairies, savannas, open woodlands, on steep slopes

BLOOM Spring, summer

DESCRIPTION Erect perennial herb, unbranched, hairy, colonial, to 15 in. tall

FLOWER Yellow, burnt orange, or reddish purple, to 1 in. long, tubular, two-lipped, the upper curved and hoodlike, the lower three-lobed, on a dense spike

LEAF Mostly basal, lance-shaped, lobed and toothed, crinkled, fernlike, alternate, to 6 in. long and 2 in. wide

FRUIT Dry, smooth capsule

Also known as lousewort for the unfounded belief that livestock eating the plant would become infested with lice. The plant is partially parasitic, drawing nutrition from roots of other plants but not totally relying on them. Reported as an Ojibwe "love charm," its use is said to turn quarrelsome couples into lovers.

Oxalis stricta
Oxalidaceae

yellow wood sorrel

HABITAT Woods, fields, roadsides, gardens, disturbed sites

BLOOM Summer, fall

DESCRIPTION Erect to ascending annual or perennial herb, branched, with spreading hairs, to 1 ft. tall

FLOWER Yellow, five-petaled, to ½ in. wide, in clusters of one to three at or above the leaves

LEAF Alternate, cloverlike, with three heart-shaped leaflets, broadest at the tip and notched, to ¾ in. long and slightly wider

FRUIT Cylindrical capsule on upright stalks, explodes forcefully to disperse seed

Also known as sour grass, *Oxalis* plants contain oxalic acid, which gives them a sour taste. In large quantities, oxalic acid can be toxic. Each night, the leaflets fold, causing their blades to assume a more vertical position. It has been suggested that this protects the wood sorrel by making animals that eat it more visible to predators.

Linaria vulgaris
Plantaginaceae

butter and eggs

HABITAT Roadsides, fields, fencerows, disturbed sites, especially with dry soil

BLOOM Summer, fall

DESCRIPTION Erect perennial herb, unbranched, hairless, colonial, to 2 ft. tall

FLOWER Yellow and orange, to 1 in. long, tubular flower faces upward with two yellow upper and three lower lobes and an orangish "bump" in the center, plus a long downward spur, several in a cluster atop the stem

LEAF Numerous, linear, entire, alternate, hairless, to 3 in. long and ¼ in. wide

FRUIT Dry, rounded capsule

This snapdragon-like plant is an invasive weed introduced from Eurasia. The common name refers to the flower color, the yellow portion being the "butter" and the orange center the "egg." Also known as yellow toadflax for the similarity of flower color and leaf shape to flax (*Linum* species). "Toad" may refer to the broad mouth of the flower. Used for making yellow dye.

Caltha palustris
Ranunculaceae

marsh marigold

HABITAT Seepage wetlands including fens, seep springs, marshes, springs

BLOOM Spring

DESCRIPTION Erect to sprawling, hairless perennial, clumped, branched above, to 2 ft. tall

FLOWER Yellow, with five or six petal-like sepals, to 2 in. wide, in branched clusters atop the stem

LEAF Rounded to kidney-shaped, alternate on stem and basal, shallowly toothed, to 6 in. long

FRUIT Pod curves out and downward at maturity, resembles a jester's hat

Marsh marigold is one of the earliest blooming wildflowers in wetland habitats. Contrary to its common name, it is not a marigold or related. It is a member of the buttercup family, and its flower most closely resembles a buttercup (*Ranunculus* species). Also known as cowslip.

Ranunculus hispidus
Ranunculaceae

hispid buttercup

HABITAT Muddy floodplain forests, marshes, swamps, wet fields, meadows, thickets, ditches, riverbanks

BLOOM Spring, summer

DESCRIPTION Erect to trailing perennial herb, stems arching and some rooting at the tips, forming colonies, hairy, to 2 ft. tall

FLOWER Yellow, five-petaled, rounded, shiny, to 1¼ in. wide, one per stalk from upper leaf axils

LEAF Compound with three leaflets, each lobed or toothed, variously hairy, oval in outline, alternate, to 5 in. long and wide

FRUIT Flat nutlet with small beak, several packed in a cluster on a central receptacle

Generally, most buttercup plants like "wet feet," living in the same sorts of places that frogs prefer. Thus, what better genus name than *Ranunculus*, which, in Latin, means "little frog." There are two varieties of hispid buttercup in the state, the differences based mostly on characters of the flowers.

Verbascum thapsus
Scrophulariaceae

common mullein

HABITAT Various disturbed sites, roadsides, railroads, fields, waste areas, ditches, pastures

BLOOM Summer

DESCRIPTION Erect biennial herb, unbranched (except sometimes in flowering stalk), hairy, to 6 ft. tall

FLOWER Yellow, to 1¼ in. wide, short-tubular, five-lobed, the lower three slightly longer, numerous, in tall, spike-like cluster to 2 ft. long

LEAF Oblong, unlobed, blunt-tipped, basal and alternate and clasping stem, densely hairy and flannel-like, to 1 ft. or more long

FRUIT Round capsules, each reported to contain 500–800 seeds (may have as many as 240,000 seeds per plant!)

The fuzzy, light green leaves of mullein are thick and soft to the touch, like flannel or felt, and have been used for shoe inserts. Introduced from Eurasia. It is said that early Greeks and Romans coated the dried flowering stalks with fat and burned them as torches. When branched, some flowering stalks resemble the shape of a saguaro cactus.

Viola pubescens
Violaceae

downy yellow violet

HABITAT Moist to dryish forests, ravines, slopes, stream terraces, thickets

BLOOM Spring

DESCRIPTION Erect perennial herb, unbranched, hairy or not, to 15 in. tall

FLOWER Yellow, to ¾ in. wide, five-petaled, with two upper petals, two side petals with hairy bases, one lower petal with short spur at base, all (most prominently the lower petal) with purplish veins, solitary on stalks from leaf axils

LEAF Heart-shaped, one or more basal and one to four alternately on stem, toothed, hairy or not, to 4 in. long and almost as wide

FRUIT Cylindrical capsule, hairless or wooly, with numerous seeds

Two versions of this violet are treated as varieties or separate species. One is mostly hairless with three or more stem leaves and one or more basal leaves (known as *V. eriocarpa*, smooth yellow violet, by some); the other is hairy with two stem leaves and no or one basal leaf.

Green Flowers

Arisaema triphyllum
Araceae

jack-in-the-pulpit

HABITAT Rich, moist forests

BLOOM Spring

DESCRIPTION Erect hairless perennial herb to 2½ ft. tall with a hooded, vase-shaped tube (spathe), to 4 in. tall, containing a single clublike structure (spadix)

FLOWER Tiny, petalless, numerous at base of club, the undersurface of the tube's hood flap either green or with dark purple stripes

LEAF One or two, divided into three leaflets to 8 in. long and 3 in. wide, with prominent veins

FRUIT Berries, several in a cluster, brilliantly red when ripe

A single jack-in-the-pulpit plant has the unusual trait of producing in any given year either male or female flowers, or both. It belongs to a mostly tropical family that includes several popular houseplants such as heartleaf philodendron and pothos, but also a few native Minnesota wildflowers such as wild calla and skunk cabbage.

Ambrosia artemisiifolia
Asteraceae

common ragweed

HABITAT Fields, croplands, roadsides, disturbed sites

BLOOM Summer, fall

DESCRIPTION Erect, branched, annual herb, variously hairy, to 3 ft. tall

FLOWER Yellowish green, tiny, in heads about ⅛ in. wide, male and female flowers occur separately

LEAF In opposite pairs on lower stem, upper ones commonly alternate, deeply dissected into narrow lobes, to 6 in. long, 4 in. wide at base

FRUIT Tiny, round bur enclosing seed

Ragweed is wind-pollinated and its airborne pollen is often a cause of hay fever for many people. Although the plant is much maligned, its seeds are important food for wildlife, especially for cardinals, chickadees, and titmice. Leaves of the similar but less common western ragweed (*A. psilostachya*) are typically thicker, hairier, and less divided.

Polygonatum biflorum
Convallariaceae

smooth Solomon's seal

HABITAT Rich, moist forests, dryish woods, thickets

BLOOM Spring, summer

DESCRIPTION Erect to arching, unbranched perennial, hairless, to 5 ft. tall (but usually much less)

FLOWER Greenish white, tubular, with five slightly flaring lobes at tip, to ¾ in. long, two to five flowers in groups dangling from leaf axils

LEAF Oval, positioned somewhat on the same plane and clasping at base, hairless, alternate, numerous parallel veins visible, to 6 in. long and half as wide

FRUIT Berry to ⅜ in. wide, deep blue when ripe

Polygonatum is from the Greek *polys* (many) and *gonu* (bended knee), for the numerous joints of the rhizome. A popular explanation for its common name refers to the rhizome's leaf scars, like wax seals impressed by King Solomon's signet ring. Also in Minnesota is hairy Solomon's seal (*P. pubescens*), with finely hairy veins on the undersurfaces of the leaf blades.

Corallorhiza trifida
Orchidaceae

early coralroot

HABITAT Moist forests, cedar swamps, thickets

BLOOM Spring, summer

DESCRIPTION Erect, unbranched perennial, hairless, to 1 ft. tall

FLOWER Green (with white and usually unspotted lowermost lip petal), three petals and three petal-like sepals, to ½ in. wide

LEAF Appearing leafless, reduced to small stem scales

FRUIT Rounded capsules, each with numerous dustlike seeds

Early coralroot is found primarily in the northern half of the state. Its rootstock has the shape of certain coral formations, hence the common and genus names. All of the four Minnesota coralroot species share the common trait of acquiring nutrition from fungi that procure it from photosynthetic plants.

Laportea canadensis
Urticaceae

wood nettle

HABITAT Floodplain forests, streambanks, ravines

BLOOM Summer

DESCRIPTION Erect perennial herb with a zigzagging stem covered with copious stinging hairs, colonial, to 3 ft. tall

FLOWER Green, about ⅛ in. wide, petals lacking, male and female separate—males on spreading branches from lower leaf axils, females from upper axils

LEAF Broadly oval with small teeth along margin, alternate, to 8 in. long and 5 in. wide

FRUIT Shiny, flat, black nutlet

Skin contact with the stinging hairs of wood nettle brings about an immediate painful reaction. Some claim relief by application of the juicy sap of jewelweed/touch-me-not (*Impatiens* species). Fortunately, the pain is usually not long-lasting, whether treated or not. The Ojibwe have made tea from it for its diuretic properties, aiming to cure various urinary ailments.

Blue to Violet Flowers

Cichorium intybus
Asteraceae

chicory

HABITAT Roadsides, waste places, fields, fencerows, disturbed sites

BLOOM Summer

DESCRIPTION Erect or ascending perennial herb, branched, mostly hairless on upper stem, hairy on lower stem, milky sap, to 3 ft. tall

FLOWER Blue, with up to 30 ray flowers, disk flowers absent, crowded in a composite head, at tips of spreading spike-like branches, to 2 in. wide

LEAF Basal, also alternate on stem, with deeply triangular lobes and irregular teeth, clasping, broadest toward tip, resembles dandelion leaf, to 12 in. long and 3 in. wide

FRUIT Dark, five-angled nutlet

Introduced from Europe, this roadside weed has flowers that open early in the morning and usually close by midday, or sometimes later if cool and cloudy. Its roots are used for tea or added to coffee or other beverages. Its foliage is edible; some well-known domesticated chicories are radicchio and Belgian endive.

Lygodesmia juncea
Asteraceae

skeletonweed

HABITAT	Dry prairies, roadsides, railroads, fields
BLOOM	Summer
DESCRIPTION	Erect perennial herb, branched, hairless, with milky sap, to 2 ft. tall
FLOWER	Violet to pinkish, usually with five ray flowers, disk flowers absent, in a composite head, solitary at branch tips, to ¾ in. wide
LEAF	Mostly inconspicuous, scalelike, alternate, lower leaves to 2 in. long and ⅛ in. wide
FRUIT	Small, dry nutlet with silky hairs that facilitate wind dispersal

Named skeletonweed because of its seemingly leafless stems, this mostly Great Plains plant occurs primarily in the western half of the state. It is commonly afflicted with pealike galls caused by a type of cynipid wasp (*Antistrophus lygodesmiaepisum*) that lays its eggs only on the stems of skeletonweed. Its larvae feed on the tissues of the galls.

Symphyotrichum novae-angliae
Asteraceae

New England aster

HABITAT Prairies, fields, meadows, lakesides, roadsides, fens, thickets

BLOOM Summer, fall

DESCRIPTION Erect perennial herb, branched above, hairy, to 5 ft. tall

FLOWER Purple, to 1½ in. wide, composed of an outer ring of up to 50 ray flowers and a similar number of yellow tubular flowers, flower stalks and head glandular, in a spreading cluster atop stem

LEAF Lance-shaped, unlobed, mostly toothless, hairy, base clasping, alternate, to 4 in. long and 1 in. wide

FRUIT Brown, narrowly wedge-shaped nutlet (seed within), tuft of hair at one end

The showy New England aster is one of the most cultivated of the native asters, with more than 50 cultivars available in the nursery trade. Another blue-flowered aster is swamp aster (*S. puniceum*). Its leaves usually have some teeth and the flower heads lack sticky glands. As its common name implies, it prefers wet environments.

Vernonia fasciculata
Asteraceae

prairie ironweed

HABITAT Wet to moist prairies, marshes, pastures, meadows, lake borders

BLOOM Summer, fall

DESCRIPTION Erect perennial herb, unbranched, hairless, to 5 ft. tall

FLOWER Purple, with up to 30 disk flowers, ray flowers absent, all crowded in numerous composite heads to ¾ in. wide, clustered in a somewhat flat-topped arrangement

LEAF Narrowly lance-shaped with sharply toothed margins, the teeth white-tipped, undersurface mostly hairless with scattered shiny resinous pits, to 6 in. long and 1½ in. wide, alternate

FRUIT One-seeded nutlets with a plume of brownish orange bristles

Some say ironweed is so named because of its tough stem, but others claim the grouping of its ripe nutlets bearing brownish orange plumes suggest rusted iron. Butterflies seek out this plant, especially in late summer and fall when fewer wildflowers are in bloom. Some members of the genus show promise as a source of antitumor agents.

Hydrophyllum virginianum
Boraginaceae

Virginia waterleaf

HABITAT Moist upland and lowland forests, ravines, floodplains, thickets

BLOOM Spring

DESCRIPTION Erect perennial herb, hairless or short-hairy, to 2½ ft. tall, rhizomatous, forming colonies

FLOWER Lavender to white, bell-shaped, five-lobed, to ½ in. wide, several in rounded clusters atop stem

LEAF Alternate, deeply lobed or compound, slightly longer than wide with three to five lobes, margins toothed, surface often with silvery white splotches, especially on the early-developed leaves, to 6 in. long

FRUIT Rounded capsule with one to three seeds

The common and genus names refer to the silvery white splotches on some leaves. In Greek, *hydrophyllum* literally means "water leaf"—*hydor* (water) and *phyllon* (leaf). Several sources state that the leaves can be cooked as a pot herb. One of the species' common names is Shawnee salad. The Ojibwe have used the root to feed ponies to rapidly fatten them.

Campanula rotundifolia
Campanulaceae

harebell

HABITAT Open woodlands, bluffs, rock outcrops, prairies

BLOOM Summer, fall

DESCRIPTION Erect perennial herb, slender, mostly hairless and unbranched, milky sap, to 2 ft. tall

FLOWER Blue, occasionally white or pink, bell- or cup-shaped, five-lobed, about ¾ in. long, nodding

LEAF Basal leaves rounded, often somewhat heart-shaped, usually absent during flowering; stem leaves linear and alternate, mostly hairless

FRUIT Capsule nodding, to ⅓ in. long, opening by pores near base where seeds are released

Also known as bluebell, this wildflower occurs throughout the Northern Hemisphere. It is the national flower of Sweden. In North America, it occurs from coast to coast, mostly in northern latitudes and in the western mountains as far south as northern Mexico. The Ojibwe name for harebell is *adota'gons* (little bell), believed to help alleviate lung troubles.

Lobelia siphilitica
Campanulaceae

great blue lobelia

HABITAT Floodplains, wet prairies, meadows, shores, streambanks, ditches

BLOOM Summer, fall

DESCRIPTION Erect perennial herb, unbranched, stem variously hairy and ribbed, to 3 ft. tall

FLOWER Blue, tubular, five-lobed, three lobes lowermost, commonly whitish near the throat, to 1 in. wide, numerous atop tall, cylindrical stalk

LEAF Elliptic to lance-shaped with small, shallow teeth, alternate, to 6 in. long and 2½ in. wide

FRUIT Capsule, round, ribbed

As indicated by its species epithet, great blue lobelia was formerly considered a treatment for syphilis. For some Indigenous tribes, it was a "love medicine," whereby roots eaten by a couple would prevent separation and renew love. Pale-spiked lobelia (*L. spicata*) has much smaller flowers that can be pale blue (or white). It is primarily a grassland species.

Tradescantia bracteata
Commelinaceae

long-bracted spiderwort

HABITAT Dry to moist prairies, savannas, roadsides

BLOOM Spring, summer

DESCRIPTION Erect perennial herb, unbranched, mostly sparsely hairy, to 1½ ft. tall

FLOWER Purple, to about 1½ in. wide, with three rounded petals and three smaller green sepals with long hairs tipped with sticky glands, six hairy stalked stamens, clustered atop the stem, each flower fresh for only one day

LEAF Linear, folded into "V" lengthwise with a sheathing base, alternate, hairless or with scattered hairs, to 8 in. long and 1 in. wide

FRUIT Capsule, somewhat egg-shaped, holds three to six seeds

The name spiderwort may refer to the long, threadlike hairs spreading from the stamen filaments. Another possibility alludes to the cobwebby appearance of dried secretions from its cut stem. Also, its long, narrow leaves arch downward, like spider legs. Also purported to be an antidote to spider bites.

Amorpha canescens
Fabaceae

leadplant

HABITAT Dry to moist prairies and savannas, woodlands

BLOOM Summer

DESCRIPTION Erect, hairy, multi-stemmed shrub to 3½ ft. tall

FLOWER Purple, consisting of a single petal ¼ in. long, numerous on mostly upright spikes to 6 in. long, normally several spreading at the tips of branches

LEAF Pinnately compound, alternate, to 12 in. long, hairy, with up to 51 grayish leaflets

FRUIT Tiny pod containing a single seed

In the wild, the presence of leadplant is usually indicative of a high-quality natural area. Its common name refers to its overall grayish appearance, thought previously by some to indicate the presence of lead ore (which is untrue). Very ornamental and worthy of cultivation, it attracts a variety of pollinators.

Astragalus crassicarpus
Fabaceae

ground plum

HABITAT Dry prairies, often on slopes

BLOOM Spring

DESCRIPTION Trailing herb with stems more or less prostrate or sprawling, hairy, to 2 ft. long

FLOWER Pinkish purple to blue-violet, five-petaled, in a typical pea family shape, to ¾ in. long, the upper petal broad, notched at tip

LEAF Compound, alternate, to 5 in. long, with up to 29 leaflets

FRUIT Round, 1 in. wide pod, shiny, fleshy, and purple to reddish when ripe, superficially resembling a miniature plum

Also called buffalo bean, its young seedpods (ground plums) are usually located on the ground and are reported to be edible, ranging in taste from apple- to pealike. Caution is advised and accurate identification is crucial before consuming them, as some similar-looking plants are considered toxic.

Dalea purpurea
Fabaceae

purple prairie clover

HABITAT Dry to moist prairies with a variety of soil types, savannas, open woodlands

BLOOM Summer

DESCRIPTION Slender, erect perennial, hairless, to 3 ft. tall

FLOWER Purple, five-petaled, to ¼ in. wide, numerous, arranged in a ring on a cylindrical thimble-like head, blooming first at bottom, progressing upward

LEAF Pinnately compound, alternate, with three to seven linear leaflets, dark green and may be somewhat shiny, to 5 in. long

FRUIT Tiny pod, densely hairy at apex, with one or two seeds

Like many members of the pea family, this widespread and distinctive prairie plant fixes nitrogen in the soil. It is also beneficial to a large variety of insects that are attracted to its pollen and nectar, including plasterer bees, miner bees, bumblebees, butterflies, and beetles. Formerly known botanically as *Petalostemum purpureum*.

Desmodium canadense
Fabaceae

showy tick-trefoil

HABITAT Prairies, fields, savannas, woodlands, thickets

BLOOM Summer, fall

DESCRIPTION Erect perennial herb, hairy, to 6 ft. tall

FLOWER Pinkish purple, with typical pea family shape, to ½ in. wide, the larger, uppermost banner petal with two lighter-colored spots outlined in purple (nectar guides) at its base, numerous in clusters atop plant

LEAF Compound, alternate with three leaflets, each to about 3 in. long, hairy

FRUIT Flat, segmented pealike pod with hooked hairs

Anyone walking in the fall through vegetation that includes this plant will likely become covered with its fruit. It's one of several "sticktights" that cling to clothing, shoelaces, and hair. Several wildflowers use this method of seed dispersal, which enables them to spread and establish populations in new locations.

Lathyrus venosus
Fabaceae

veiny pea

HABITAT Dry, open woodlands, savannas, prairies, thickets

BLOOM Summer

DESCRIPTION Sprawling, climbing perennial herb, stem four-angled and mostly hairless, often in colonies, to 3 ft. tall

FLOWER Blue to purple, of typical pea family shape, to ¾ in. long, two-toned, the uppermost petal darker than the others and with dark veins, in dense clusters of up to 20 flowers

LEAF Pinnately compound with up to 12 oval leaflets, to 8 in. long, each leaf tip bearing a coiled tendril instead of a leaflet

FRUIT Narrow, flat pod, to 3 in. long

Veiny pea is the most widespread of the upland growing wild peas in Minnesota. Cream pea (*L. ochroleucus*) has whitish flowers and marsh pea (*L. palustris*) has fewer leaflets per leaf and occurs in wet habitats. *Lathyrus* may be derived from *la* (very) and *thuros* (passionate). The original-named plant of the genus, from Europe, was thought to be an aphrodisiac.

Pediomelum esculentum
Fabaceae

prairie turnip

HABITAT Dry prairies, savannas

BLOOM Spring, summer

DESCRIPTION Erect perennial herb, branched, hairy, from a thickened taproot, to 16 in. tall

FLOWER Purple, with a typical pea-flower shape, to ¾ in. long, five-petaled, in a crowded leafy spike atop stem

LEAF Compound, alternate, with five leaflets palmately arranged, hairy, to 2½ in. long and ½ in. wide

FRUIT One-seeded pod to ¼ in. long

This member of the pea family is mostly restricted to the western half of the state, where there is a greater occurrence of prairie. It has a starchy rootstock that has been an important food for Native Americans. In some Plains tribes it's known as *timpsila*, and the roots are braided together and dried for later use.

Gentiana andrewsii
Gentianaceae

bottle gentian

HABITAT Wet meadows, prairies, marshes, shores, open woodlands

BLOOM Late summer, fall

DESCRIPTION Erect to somewhat lax, usually unbranched perennial herb, essentially hairless, to 2½ ft. tall

FLOWER Blue, tubular, with five vertically oriented floral lobes that overlap, closed except for a tiny opening at the tip, to 1½ in. long, somewhat bottle-shaped, several clustered atop stem

LEAF Broadly lance-shaped, glossy, unlobed, opposite, to 4 in. long and about as wide

FRUIT Capsule, with numerous papery winged seeds

Bottle gentian, also known as closed gentian, is one of the last wildflowers of the year to bloom. Overlapping floral lobes present a challenge to potential pollinators, but large insects such as bumblebees are up to the task, strong enough to force their way through the top of the flower. Flowers can be pinkish or white, but blue is the most common color.

Geranium maculatum
Geraniaceae

wild geranium

HABITAT Moist forests, forest edges, stream terraces, thickets

BLOOM Spring

DESCRIPTION Erect perennial herb, hairy, unbranched, usually clumped, to 2½ ft. tall

FLOWER Pinkish purple, with five broadly rounded petals, to 1½ in. wide, in loose clusters atop stem

LEAF Basal leaves and oppositely attached pairs on stem palmately divided, these further divided into broad teeth, to 5 in. long and about as wide

FRUIT Narrow, beaklike; when mature, outer covering splits into five separate strips that curl upward from the base, slinging seeds some distance away

Unlike commercial potted geraniums, our wild geranium is the real deal. Potted selections are of the genus *Pelargonium*, not *Geranium*. One common name for the group is cranesbill, a name referencing the long, narrow shape of the fruit suggesting the beak of a crane. The name *Geranium* is from the Greek *geranos*, which means "crane."

Iris versicolor
Iridaceae

northern blue flag

HABITAT Marshes, wet meadows, swamps, bogs, lakeshores, streams, ditches

BLOOM Summer

DESCRIPTION Erect perennial herb, hairless, colonial, from horizontal rhizomes

FLOWER Blue to violet, to 4 in. wide, parts in threes, with three petal-like outer sepals that curve downward, each bearing a greenish yellow splotch near its base, and three inner upright petals, the stalked flowers usually reach above the leaf tips

LEAF Stiff or arching, swordlike blades to 3 ft. tall, growing from horizontal rhizomes

FRUIT Three-angled, cylindrical capsule

This plant is named after Iris, the Greek goddess of rainbows. The blue flag common name likely refers to its leaves that flutter like flags in the wind. Flowers of southern blue flag (*I. virginica*) occur below the leaf tips and have sepals with a brighter yellow splotch. It occurs mostly south of Minneapolis.

Agastache foeniculum
Lamiaceae

blue giant hyssop

HABITAT Thickets, woodlands, fields

BLOOM Summer

DESCRIPTION Erect, usually unbranched herb with a square stem, to 3 ft. tall, smooth or with few hairs

FLOWER Blue, to ½ in. long, tubular, with long exerted stamens, two-lipped, the upper two-lobed, lower three-lobed, densely packed into spikes

LEAF Lance- or egg-shaped, opposite, to 4 in. long, coarsely toothed, undersurface matted with short, grayish hairs

FRUIT Vase-shaped, dry, typically with four nutlets

Also known as anise hyssop for the anise scent of its crushed foliage. Its flowers attract several species of bees and other insects and is sometimes used in home landscaping. Purple giant hyssop (*A. scrophulariifolia*) is similar, but its leaves are green underneath and the species is mostly confined to the southeastern counties.

Prunella vulgaris
Lamiaceae

self-heal

HABITAT Open woodlands, trails, roadsides, clearings, fields, meadows, disturbed sites

BLOOM Summer

DESCRIPTION Erect or prostrate perennial herb, mostly unbranched, hairy or not, stems four-angled, to 18 in. tall

FLOWER Bluish purple, ½ in. long, tubular, two-lipped, the upper one purplish and hoodlike, the lower lip three-lobed with the lowermost lobe often white, several in a dense cylindrical spike atop the stem

LEAF Lance-shaped to elliptic, margins may have a few shallow teeth, to 3 in. long and about half as wide, opposite

FRUIT One-seeded nutlets

Two varieties of self-heal exist, one or perhaps neither native to North America. It has a long history of use as folk medicine, believed to cure an assortment of ailments. Although in the mint family, it lacks a minty scent. The Ojibwe have made tea from its roots, said to sharpen their powers of observation for hunting.

Scutellaria galericulata
Lamiaceae

marsh skullcap

HABITAT Marshes, swamps, wet meadows, fens, bogs, thickets, shores, ditches

BLOOM Summer

DESCRIPTION Sprawling, weak-stemmed perennial, stems square with hairs on angles, branched, to 2½ ft. tall

FLOWER Blue with lightly spotted white throat, tubular, to 1 in. long, two-lipped, upper forming a hood over the lower flattened lip, finely hairy, occurring singly from axils of leaves facing in same direction and appearing as pairs

LEAF Lance-shaped, unlobed, with blunt teeth along margins, deeply veined, short-stalked, progressively smaller upward, to 2¼ in. long and ¾ in. wide

FRUIT Rounded nutlets in a fruiting structure that resembles an old-fashioned tractor seat

Marsh skullcap occurs naturally in North America, Europe, and Asia. Of our Minnesota species, it is most like mad-dog skullcap (*S. lateriflora*), whose flowers are almost half the size and lack spots on the lower lip. Both occur in wet habitats. The Ojibwe have used marsh skullcap to treat heart troubles.

Lythrum salicaria
Lythraceae

purple loosestrife

HABITAT Open wetlands, marshes, meadows, shores, streams, ditches

BLOOM Summer, fall

DESCRIPTION Erect perennial herb, branched (even bushy), square-stemmed, hairy, to 6 ft. tall

FLOWER Purplish, to 1 in. wide, six-petaled, often with a wrinkled appearance, in dense groups on tall, narrow spikes atop the stems

LEAF Lance-shaped, variously hairy, opposite and somewhat clasping the stem, some alternate or whorled, to 4 in. long and 1 in. wide

FRUIT Dry, egg-shaped capsule to ⅛ in. long

Despite their common names, purple loosestrife and tufted loosestrife are not related. Purple loosestrife is Eurasian in origin and extremely invasive. It has negative impacts on native species and can displace them. Efforts have been made to reduce its numbers using insects from its homeland that specialize in feeding on it. So far it appears to be working.

a different "fireweed"
(*Erechtites hieraciifolius*)

Chamaenerion angustifolium
Onagraceae

fireweed

HABITAT Moist clearings, roadsides, ditches, recently burned or disturbed sites

BLOOM Summer

DESCRIPTION Erect perennial herb to 7 ft. tall, colonial by rhizomes

FLOWER Purplish pink, four-petaled, oval-shaped, to 1 in. wide, in long cluster at stem tip

LEAF Narrow, spear-shaped, edges wavy or slightly toothed, alternate, to 8 in. long

FRUIT Slender, four-angled capsule, splits like banana peel revealing seeds, each bearing a long plume of hairs at tip

Fireweed is so named because of its common occurrence in recently burned habitats. At a distance, it may be confused with the similar-looking purple loosestrife (*Lythrum salicaria*), but flowers of that highly invasive, nonnative weed of wetlands are six-petaled. In eastern Minnesota, another plant called fireweed (*Erechtites hieraciifolius*, see inset), has a head of tiny white flowers.

Platanthera psycodes
Orchidaceae

lesser purple fringed orchid

HABITAT Seepage wetlands, swamps, marshes, streambanks

BLOOM Summer

DESCRIPTION Erect perennial herb, unbranched, hairless, to 3 ft. tall

FLOWER Purple, fan-shaped, to ¾ in. long, with three petal-like sepals and three petals, one of which is larger and lowermost with three deeply fringed lobes and a long spur at base, clustered on a stalk atop the stem

LEAF Lance-shaped, unlobed, hairless, alternate, to 7 in. long and 3 in. wide

FRUIT Capsule with numerous dustlike seeds

This wild orchid is one of 11 species of rein orchids (*Platanthera* species) in the state. They are collectively called rein orchids because they have nectar spurs that fancifully resemble the reins (straps) of a horse's bridle. In Minnesota, lesser purple fringed orchid is known mostly from the eastern third of the state.

Mimulus ringens
Phrymaceae

Allegheny monkeyflower

HABITAT Marshes, fens, shores, swamps, streams, ditches

BLOOM Summer

DESCRIPTION Erect perennial herb, square-stemmed, hairless, to 3 ft. tall

FLOWER Blue with a yellowish center, to 1 in. long, two-lipped with wavy edges, the upper two-lobed, the lower three-lobed and widely spreading, on 1 to 2 in. stalks emerging from leaf axils

LEAF Lance-shaped with small teeth on margins, opposite, clasping at base, to 4 in. long and 1 in. wide

FRUIT Elliptic-shaped capsules

Known as monkeyflower perhaps because of the fanciful resemblance of the flower to that of an expressive, "smiling" face of a monkey or someone monkeying (or clowning) around. Such a look is also reflected in its genus name *Mimulus*, from *mimus*, for "clown." Its large, showy flowers are frequented by bumblebees and the species is often used in pollination research.

Penstemon grandiflorus
Plantaginaceae

large beardtongue

HABITAT Dry prairies, savannas, barrens, especially in areas of sand and rock deposits

BLOOM Spring, summer

DESCRIPTION Erect perennial herb, unbranched, hairless, with a bluish cast, to 3½ ft. tall

FLOWER Purplish pink, to 2 in. long, tubular, upper lip two-lobed, lower lip three-lobed, flared at the opening, in groups of two to six spaced along stalk at top of stem

LEAF Rounded, thick and waxy, basal, opposite on upright stem, hairless, to 5 in. long, smaller on stem

FRUIT Teardrop-shaped capsule, to 1 in. long

Large, showy flowers characterize this plant, which is grown as an ornamental in some home gardens. The genus name *Penstemon*, of which there are a handful of species in the state, means “five stamens,” one of which is sterile. Many of the species sport a “beard” of hairs on the sterile stamen, giving rise to the common name beardtongue.

Anemone patens
Ranunculaceae

American pasqueflower

HABITAT Dry prairies, especially on substrates of sand or gravel

BLOOM Spring

DESCRIPTION Erect, unbranched perennial, hairy, to 15 in. tall

FLOWER Purple to white, to 2½ in. wide, occurring singly atop stem, five to seven sepals are petal-like

LEAF Deeply divided into narrow lobes, basal leaves stalked, stem leaves unstalked, barely developed when flowering occurs

FRUIT Seedlike fruits harbor threadlike extensions abundantly covered with hairs, giving a feathery look

In the prairie, there is perhaps no better harbinger of spring than pasqueflower. Depending on the part of the state, some of these beauties arise from the recently frozen earth and bloom in mid-April, others possibly earlier where exposed to direct rays of the sun, such as south-facing hillsides. *Pulsatilla patens* is another scientific name for this species.

Verbena hastata
Verbenaceae

blue vervain

HABITAT Wet prairies, meadows, marshes, shores, ditches

BLOOM Summer, fall

DESCRIPTION Erect perennial herb, unbranched except in flower clusters, angled, somewhat hairy, to 5 ft. tall

FLOWER Blue to pinkish purple, to ¼ in. wide, tubular with five spreading lobes, numerous on dense spikes to 5 in. long at tips of branches atop the plant, somewhat candelabra-like

LEAF Narrow, lance-shaped, sharply toothed with deep veins on the blade surface, opposite, often with smaller leaves in axils of main leaves, to 6 in. long and 1 in. wide

FRUIT Small nutlets, usually in groups of four within flower remnants

This species makes a nice addition to rain gardens. The verbena bee (*Calliopsis verbenae*) specializes in acquiring food from the flowers of vervains. Vervains also serve as host plants for feeding caterpillars of the verbena moth (*Crambodes talidiformis*).

Viola sororia
Violaceae

woolly blue violet

HABITAT Upland forests, savannas, fields, meadows, roadsides, thickets

BLOOM Spring

DESCRIPTION Erect or ascending perennial herb, unbranched, hairy, stemless (leaves and flowering stalks come directly from the rhizome), to 6 in. tall

FLOWER Blue with a whitened throat, to ¾ in. wide, five-petaled, with two upper petals, two side petals with hairs at their bases, and one lower petal with a short spur at its base, all (but most prominently the lower petal) with purplish veins, solitary on stalks from rhizome

LEAF Heart-shaped, basal, toothed, hairy, to 3 in. long and about as wide

FRUIT Cylindrical capsule, hairless, with numerous seeds

The similar common blue violet (*V. communis*) is typically hairless and more likely to be found in disturbed sites. About 20 species of violets occur naturally in Minnesota, including some with white and yellow flowers. Fritillary butterfly caterpillars, including the magnificent regal fritillary, feed exclusively on leaves of violet species.

Brown to Maroon Flowers

leaves

Symplocarpus foetidus
Araceae

skunk cabbage

HABITAT Seepage swamps and springs, fens, bogs

BLOOM Late winter, spring

DESCRIPTION Erect perennial herb, unbranched, hairless, colonial, to 2 ft. tall

FLOWER Maroon (hood), four tepals, several packed tightly together on a club-shaped structure (spadix), housed within a tepee-shaped, leathery hood (spathe)

LEAF Heart-shaped or oval, basal, unlobed, toothless and hairless, to 2 ft. long and half as wide

FRUIT Knobby compound fruit to 4 in. long and 3 in. wide, shape suggestive of a small pineapple

Found mostly in the eastern counties, this is our earliest-flowering native plant, often blooming in late winter. Perhaps its most unusual characteristic is its flowering spadix (club), capable of generating heat considerably warmer than the surrounding air. This trait can even melt snow on and around its hood during blooming.

Asarum canadense
Aristolochiaceae

wild ginger

HABITAT Rich, moist forests, ravines, stream terraces

BLOOM Spring

DESCRIPTION Low-growing perennial herb, unbranched, colonial, to 6 in. tall

FLOWER Maroon, tubular or urn-shaped, with three long-tapering, petal-like sepals (petals absent) often curved backward, hairy on the outside, occurring singly between paired stems at ground level

LEAF Heart- to kidney-shaped, sparsely hairy, paired at stem tip, from creeping stem along ground, to 3 in. long and 5 in. wide

FRUIT Fleshy, rounded capsule

The flower of wild ginger is hidden beneath the leaves. It is scentless and, contrary to what is often stated, likely not pollinated by carrion-seeking insects. Research indicates it is mostly self-pollinating. The common name alludes to the rhizomes that smell remarkably like that of tropical ginger (*Zingiber officinale*) used in cooking.

Sarracenia purpurea
Sarraceniaceae

purple pitcher plant

HABITAT Bogs, fens, swamps

BLOOM Summer

DESCRIPTION Erect to sprawling perennial herb, essentially stemless other than flowering stalks, clumped, carnivorous, to 2 ft. tall

FLOWER Maroon to purple, nodding, to 2½ in. wide, five-petaled, thin and deciduous, five thick sepals persistent on umbrella-like disk

LEAF Basal, tubular, filled with rainwater, rimmed on one side by an erect hood, hairless on the outside, inside covered with downward-pointing hairs, to 8 in. long

FRUIT Round to oval capsule

Pitcher plants absorb nutrients from organisms trapped in the "pools of death" in their leaves. The usual victims are insects but can include salamanders. The pitcher plant mosquito (*Wyeomyia smithii*) lays its eggs only in pitcher plant pools. Amazingly, in our area, these mosquitos do not feed on humans! An Ojibwe name for the plant is *o'makaki'wîdass*, meaning "frog's leggings."

flowers

Scrophularia lanceolata
Scrophulariaceae

early figwort

HABITAT Open woodlands, forest edges and clearings, thickets, roadsides, fencerows

BLOOM Summer

DESCRIPTION Erect perennial herb, square-stemmed, minutely hairy, to 6 ft. tall

FLOWER Reddish brown, goblet-shaped, to ⅓ in. long, two-lipped, the upper two-lobed, the lower three-lobed with two lateral lobes upright and one lower lobe bent down, interior with a yellow or green, flattened sterile stamen, stalks with tiny glandular hairs, numerous at branching stem tips

LEAF Lance-shaped, coarsely toothed margins, straight-edged or rounded base and pointed tip, opposite, to 8 in. long and 3 in. wide

FRUIT Dull brown capsule narrowed to a beak at tip

Described as one of the most prolific nectar producers in the world, the inconspicuous flowers of figworts attract hummingbirds and many insects, even bald-faced hornets. Its common name refers to its use in treating "figs," an old name for hemorrhoids. The genus name refers to its former use to treat scrofula, infected lymph nodes of the neck.

Typha latifolia
Typhaceae

broad-leaved cattail

HABITAT Marshes, lake and pond borders, ditches, fens, wetlands

BLOOM Summer

DESCRIPTION Erect perennial herb, unbranched, hairless, colonial, to 9 ft. tall

FLOWER Brownish or tan (when mature), tiny and inconspicuous, sexes separate, the tan male flowers compact on the spike directly atop the much-thicker portion containing the brown female flowers, reported to be as many as two million, to 1½ in. thick

LEAF Strap-shaped, basal and alternate, about 1 in. wide, to more or less as long as flowering stem

FRUIT Tiny nutlet with "fluff" of hairs attached to its stalk

There are two species of cattails in Minnesota, this one and the nonnative narrow-leaved cattail (*T. angustifolia*). The hybrid between the two (*T. ×glauca*), is quite invasive and forms monocultures to the detriment of native vegetation. The native cattail has long been important to Indigenous people, particularly for construction materials and food.

INTERPRETING SCIENTIFIC NAMES

A scientific name often provides information about distinctive features of a plant. Many Latin words or variations are used in the binomial names of certain species treated in this book. Consider, for example, the scientific name for red clover: *Trifolium pratense*. Even without seeing the plant, you might surmise from its genus name that it has three leaves (*tri* for "three," *folium* for "leaves"), and indeed this would be correct (actually, this species has three leaflets). Also, its specific epithet *pratense* means "of meadows," which is a common habitat for the plant.

LATIN WORD	MEANING
albidus, alba	white
alta	tall
angusti-	narrow
-anthus	flower
arvense	field
aureum	golden
bi-	two
blanda	white
boreale	northern
canescens	white-hairy
-carpum	fruit(ed)

LATIN WORD	MEANING
-caulus	stem
cernuum	nodding
cinque	five
coccineus	scarlet
cordi-	heart-shaped
crassi-	fleshy
erigeron	early, old man
flora-	flower(ed)
-folium	leaved
galericulata	covered with a helmet or hood
glaber	smooth
hirta	short-hairy
-issima	very (superlative)
lanceolatus	lance-leaf
lati-	broad
lepto-	slender, small
lithos	stone
macro-	large
maculata	spotted
mille-	thousands, many
nuda-	naked
officinalis	healing
-oides	like

LATIN WORD	MEANING
-opsis	head
palustris	marsh-loving
parva-	small
patens	spreading
pauci-	few
-phylla	leaf
platy-	broad
poly-	many
pratensis	meadow
purpureus	purple
repens	creeping
rotundi-	round, plump
rubra	red
saligna	willow-leaved
saxifraga	rock-breaker
sepium	growing in hedges
stella	star-shaped
stricta	stiff
tri-	three
umbellate	umbrella-shaped
varians	variable
villosa	downy
vulgatum	common

BIBLIOGRAPHY

Curtis, J. T. 1959. *The Vegetation of Wisconsin: An Ordination of Plant Communities*. Madison: University of Wisconsin Press.

Fernald, M. L. 1950. *Gray's Manual of Botany: A Handbook of the Flowering Plants and Ferns of the Central and Northeastern United States and Adjacent Canada*. 8th ed. New York: American Book Company.

Flora of North America Editorial Committee, eds. 1993+. *Flora of North America North of Mexico*, 20+ vols. New York and Oxford: Oxford University Press.

Gleason, H. A., and A. Cronquist. 1991. *Manual of Vascular Plants of Northeastern United States and Adjacent Canada*. 2nd ed. Bronx, NY: The New York Botanical Garden.

Homoya, M. A., and S. A. Namestnik. 2022. *Wildflowers of the Midwest*. Portland, OR: Timber Press.

Ladd. D. 1995. *Tallgrass Prairie Wildflowers: A Field Guide*. Helena, MT: Falcon Press.

Ladd, D. 2001. *North Woods Wildflowers: A Field Guide to Wildflowers of the Northeastern United States and Southeastern Canada*. Helena, MT: Falcon Press.

Smith, W. R. 2012. *Native Orchids of Minnesota*. Minneapolis: University of Minnesota Press.

Tester, J. R. 1995. *Minnesota's Natural Heritage: An Ecological Perspective*. Minneapolis: University of Minnesota Press.

Voss, E. G., and A. A. Reznicek. 2012. *Field Manual of Michigan Flora*. Ann Arbor: University of Michigan Press.

Wilhelm, G., and L. Rericha. 2017. *Flora of the Chicago Region: A Floristic and Ecological Synthesis*. Indianapolis, IN: Indiana Academy of Science.

PHOTO & ILLUSTRATION CREDITS

Map on page 14 illustrated by Michele Angel

Cover photos by Andrew Lane Gibson, Eric Hunt, Carmen Rieb/Shutterstock, and Haley Snow/Shutterstock

Flower icon by YuningArt Studio

Adam Balzer, 182

Katy Chayka, Minnesota Wildflowers, 80, 152 (inset), 214, 310, 338

Peter Dzuik, Minnesota Wildflowers, 140, 176, 248, 308

Andrew Lane Gibson, 184

Peter Grube, 180, 318

Michael Homoya, 50 (top), 60, 78 (inset), 84, 92, 96 (inset), 104, 114, 120, 134, 146 (inset), 158, 170, 172, 174, 192, 198, 224, 230, 238, 244, 250, 252, 254, 256, 262, 264, 266, 272, 290, 306, 316 (inset), 332 (bottom)

Michael Huft, 72, 118, 124, 128, 160, 274, 320, 326

Eric Hunt, 90

Scott Namestnik, 48, 52, 54, 56, 58, 62, 64, 66, 68, 70, 74, 76, 82, 86, 94, 96, 98, 100, 106, 108, 110, 112, 116, 120 (inset),122, 126, 130, 132 (inset), 136, 138, 142, 144, 146, 148, 150, 152, 156, 162, 166, 168, 186, 188, 196, 200, 202, 206, 208, 212, 216, 218, 220, 222, 226, 228, 232, 234, 236, 240, 242, 246, 258, 260, 270, 276, 280, 282, 286, 288, 292, 294, 296, 298, 302, 304, 314, 316, 322, 328, 332 (top), 336, 340

Nathanael Pilla, 178

Corey Raimond, 50 (bottom), 150 (inset), 300, 324

Paul Rothrock, 176 (inset)

Perry Scott, 88, 102, 164, 190, 210, 312

B.S. Slaughter, 284, 334

Dan Tenaglia, 78

Shutterstock

Alan B. Schroeder, 40–41

Brian Woolman, 38

BrittJPete, 132

Carmen Rieb, 20

Cynthia Shirk, 5

Darren Berendt, 8

Kevin Collison, 12

Mike Truchon, 42

torook, 2–3

WEB RESOURCES

The Biota of North America Program: bonap.org

Consortium of Midwest Herbaria: midwestherbaria.org

Flora of North America: eFloras.org

Flora of Wisconsin: wisflora.herbarium.wisc.edu

Illinois Wildflowers: illinoiswildflowers.info

Michigan Flora (University of Michigan Herbarium): michigan-flora.net

Minnesota Wildflowers: minnesotawildflowers.info

NatureServe network of Natural Heritage Programs: natureserve.org/natureserve-network/united-states

NATIVE PLANT SOCIETIES

Every state in the Midwest has one or more organizations that focus on native plants. Readers are encouraged to consult websites of several native plant–focused organizations.

North American Native Plant Society: nanps.org

Minnesota Native Plant Society: mnnps.org

The Prairie Enthusiasts: theprairieenthusiasts.org

Michigan Botanical Society: michiganbotanicalsociety.org

INDEX

Achillea, 63
Achillea millefolium, 63
Achilles, 63
Acorus americanus, 199
Acorus calamus, 199
Actaea pachypoda, 125
Actaea rubra, 125
adota' gons (little bell), 283
Agastache foeniculum, 307
Agastache scrophulariifolia, 307
Ageratina altissima, 65
agoñgosi'mînûn (chipmunk berries), 79
Allegheny monkeyflower, 319
Alliaria petiolata, 71
Allium burdickii, 47
Allium canadense, 149
Allium stellatum, 149
Allium tricoccum, 47
alpine enchanter's nightshade, 113
Ambrosia artemisiifolia, 263
Ambrosia psilostachya, 263
Ambrosia species, 215
American basswood, 13, 15
American elm, 15
American hog peanut, 95
American lopseed, 119
American lotus, 231
American pasqueflower, 323
American plum, 15
American sweetflag, 199
Amorpha canescens, 289
Amphicarpaea bracteata, 95
Andrena ziziae, 201
Anemone canadensis, 127
Anemone patens, 323
Anemone quinquefolia, 129
anise hyssop, 307
aniseroot, 55
Antennaria neglecta, 67
Anticlea, 107
Anticlea elegans, 107
Antistrophus lygodesmiaepisum, 275
Apocynum androsaemifolium, 57
Apocynum cannabinum, 57
Apocynum species, 23
apple, 33
Aquilegia canadensis, 183
Aralia nudicaulis, 61
Arisaema triphyllum, 261
arrowhead, 45
arum family (Araceae), 23, 24, 59
Asarum canadense, 333
Asclepias incarnata, 151
Asclepias species, 23

Asclepias sullivantii, 153
Asclepias syriaca, 18, 153
Asclepias tuberosa, 189
Asclepius, 151
Asian bleeding-heart, 115
aster, 277
aster family (Asteraceae), 24–25
Astragalus crassicarpus, 291
Averrhoa carambola, 32

baby's breath, 26
balsam apple, 87
balsam fir, 13
balsam ragwort, 211
baneberry, 125
Barbara, Saint, 225
Barbarea vulgaris, 225
barberry family (Berberidaceae), 25
barrens, 14, 17
bastard toadflax, 141
bean/legume family, 29, 231
bearded iris, 29
bedstraw, 139
bedstraw family, 33
bee balm, 169
bees, 169
beetles, 293
beggar-ticks, 203
Belgian endive, 273
belle of the ball, 83
bellflower family (Campanulaceae), 26
bellwort, 229
Berberis thunbergii, 25
Berteroa incana, 71
Besseya bullii, 9
Bidens cernua, 203
bigtooth aspen, 13
birds, 61, 263
bird's-foot trefoil, 231
birthwort, 145
bishop's cap, 143
black ash, 13
blackberry, 33
black cherry, 13
black-eyed Susan, 213
black spruce, 13
black swallowtail butterfly, 201
bladderwort family (Lentibulariaceae), 29
blister rust, 141
bloodroot, 117
bloom, 40
bluebeard lily, 227
bluebell, 283
blue cohosh, 25, 221
blue giant hyssop, 307
Blue Mounds State Park, 16
Bluestem Prairie Scientific and Natural Area, 16
blue to violet flowers, 271–327
blue vervain, 325
bog, 13, 17
borage family (Boraginaceae), 25
bottle gentian, 301

Bougainvillea species, 31
Boundary Waters Canoe Area Wilderness, 13
box elder, 15
Brasenia schreberi, 165
broad-leaved cattail, 339
broomrape family (Orobanchaceae), 31–32
brown to maroon flowers, 329–339
buckwheat, 181
buckwheat family (Polygonaceae), 33
buffalo bean, 291
Buffalo River State Park, 16
bûgwa'djijîca'gowûnj (unusual onion), 47
bumblebees, 293
bunch flower family (Melanthiaceae), 30
bur oak, 15
butter and eggs, 249
buttercup, 251, 253
buttercup family (Ranunculaceae), 33, 251
butterflies, 103, 279, 293
butterfly milkweed, 189

calla lily, 59
Calla palustris, 59
Calliopsis verbenae, 325
Caltha palustris, 251
Calystegia sepium, 83
camas, 107
Camassia quamash, 107
Campanula rotundifolia, 283
Canada anemone, 127
Canada mayflower, 79
Canada thistle, 155
Capsella bursa-pastoris, 73
Cardamine concatenata, 75
cardinals, 263
carnation, 26
Carolina spring beauty, 109
carrot, 23
carrot family (Apiaceae), 23, 51
Castilleja coccinea, 195
caterpillars, 153, 173, 179, 189, 201, 241, 325, 327
cattail, 339
cattail family (Typhaceae), 35
Caulophyllum thalictroides, 221
Chamaenerion angustifolium, 315
Chelone glabra, 121
chickadees, 263
chickweed, 26
chicory, 273
Chippewa National Forest, 16
Chrysochus auratus, 57
Cichorium intybus, 273
Cicuta maculata, 49
cilantro, 23
Circaea alpina, 113
Circaea canadensis, 113
Cirsium arvense, 155
Cirsium muticum, 157
Cirsium species, 157

Claytonia caroliniana, 109
Claytonia species, 30
Claytonia virginica, 109
cleavers, 139
Clematis occidentalis, 131
Clematis virginiana, 131
Clintonia borealis, 227
closed gentian, 301
clover, 167, 233, 293
Coffea species, 33
coffee, 33, 139
coffee family (Rubiaceae), 33–34
Comandra umbellata, 141
common arrowhead, 45
common bladderwort, 235
common blue violet, 327
common butterwort, 29
common dandelion, 217
common dodder, 85
common dogbane, 57
common evening primrose, 241
common milkweed, 18, 153
common mountain mint, 103
common mullein, 255
common names, 21
common ragweed, 263
common sneezeweed, 205
common sunflower-, 207
coneflower, 159
Conium maculatum, 49
Convallaria majalis, 27, 79
Corallorhiza trifida, 267
coralroot, 267
cow parsnip, 53
cowslip, 251
crabapple, 33
Crambodes talidiformis, 325
cranesbill, 303
cream pea, 297
creeping thistle, 155
Cryptotaenia canadensis, 51
Cryptotaenia japonica, 51
cucumber, 27
Culver's root, 123
Cuscuta gronovii, 85
Cuscuta species, 9
cutleaf toothwort, 75
cynipid wasp, 275
Cypripedium acaule, 175
Cypripedium parviflorum, 243
Cypripedium parviflorum var. *makasin*, 243
Cypripedium parviflorum var. *pubescens*, 243
Cypripedium reginae, 177

Dalea purpurea, 293
dandelion, 217
Darwin, Charles, 89
Delphinium carolinianum, 133
Delphinium carolinianum subsp. *virescens*, 133
descrption, 40
Desmodium canadense, 295
Dicentra canadensis, 115
Dicentra cucullaria, 115

Dicentra spectabilis, 115
dill, 23
disk flowers, 25
disturbed site, 17–18
dodder, 9
dogbane, 57
dogbane beetle, 57
dogbane family (Apocynaceae), 23–24
dolls eyes, 125
downy yellow violet, 257
Driftless Area, 15
Drosera rotundifolia, 89
duck potato, 45
Dutchman's breeches, 115
Dutchman's pipevine, 24
dwarf trout lily, 9, 105

early coralroot, 267
early figwort, 337
echinacea, 159
Echinacea angustifolia, 159
Echinocystis lobata, 87
ecoregions, 13–15
enchanter's nightshade, 113
Enemion biternatum, 129
Erechtites hieraciifolius, 315
Erigeron philadelphicus, 69
Erythronium albidum, 105
Erythronium americanum, 105
Erythronium propullans, 9, 105
Erythronium species, 30
Eupatorium rugosum, 65
Eutrochium maculatum, 161
evening primrose, 241
evening primrose family (Onagraceae), 31
Exacum affine, 28

false rue anemone, 129
fen, 13, 18
field pussytoes, 67
figwort, 337
figwort family (Scrophulariaceae), 31, 32, 34
fire, 37
fireweed, 315
flower, 40
Foot Hills State Forest, 16
forest, 13, 18
four-o'clock, 173
four-o'clock family (Nyctaginaceae), 31
Fragaria vesca, 137
Fragaria virginiana, 137
fragrant white waterlily, 111
fritillary butterfly caterpillars, 327
fruit, 40

Galium aparine, 139
Galium boreale, 139
galls, 275
garden phlox, 33
garlic mustard, 71
Gentiana andrewsii, 301
gentian family (Gentianaceae), 28

geranium, 9, 29, 303
geranium family (Geraniaceae), 29
Geranium maculatum, 303
germander, 171
German iris, 29
giant sunflower, 207
ginger, 333
ginseng, 24, 61
ginseng family (Araliaceae), 24
Glycyrrhiza glabra, 97
Glycyrrhiza lepidota, 97
glycyrrhizin, 97
goat's beard, 219
golden alexanders, 201
goldenrod, 215
gourd family (Cucurbitaceae), 27
great blue lobelia, 285
green ash, 15
green flowers, 259–269
ground plum, 291

habitat, 37, 40
hackberry, 15
hairy puccoon, 223
hairy Solomon's seal, 265
harebell, 283
heartleaf philodendron, 24, 261
heart-leaved golden alexanders, 201
heath family (Ericaceae), 29
hedge bindweed, 83
Helenium autumnale, 205
Helianthus giganteus, 207
Heliodines nyctaginella, 173
Heliopsis helianthoides, 207
Hepatica acutiloba, 135
Hepatica americana, 135
Hepatica nobilis, 135
Heracleum maximum, 53
hispid buttercup, 253
hoary alyssum, 71
hoary puccoon, 223
Ho-Chunk people, 91
honewort, 51
hummingbirds, 337
Hydrophyllum virginianum, 281

Impatiens, 25
Impatiens capensis, 191
Impatiens pallida, 191
Impatiens species, 191, 269
Indian paintbrush, 195
Indian pipe, 91
Indigenous Americans, 53, 63, 123, 145, 159, 209, 223, 285, 339
insects, 37
invasive plants, 18, 37, 155, 249, 313
iris, 9, 29
Iris, 21
Iris (Greek goddess of rainbows), 305
iris family (Iridaceae), 29
Iris germanica, 29

Iris versicolor, 21, 305
Iris virginica, 21, 305
ironweed, 279
Itasca State Park, 16

jack-in-the-pulpit, 59, 261
jack pine, 14
Japanese barberry, 25
jewelweed, 191, 269
John-go-to-bed-at-noon, 219

kittentails, 9
knotweed, 181

lady's slippers, 175, 243
lake, 13, 18
lake plains, 13
Laportea canadensis, 269
large beardtongue, 321
large-flowered bellwort, 229
large-leaved shinleaf, 93
Lathyrus ochroleucus, 297
Lathyrus palustris, 297
Lathyrus venosus, 297
leadplant, 289
leaf, 40
Leopold, Aldo, 10
lesser purple fringed orchid, 317
Liatris aspera, 163
Liatris pycnostachya, 163
Liatris species, 163
licorice, 97
Lilium michiganense, 193
Lilium philadelphicum, 193
Lilium species, 30, 193
lily, 9, 27
lily family (Liliaceae), 30, 35, 59
lily-of-the-valley, 27, 79
lily-of-the-valley family (Convallariaceae), 27
Linaria vulgaris, 249
Lincoln, Abraham, 65
Lincoln, Nancy Hanks, 65
Linum species, 249
Lithospermum canescens, 223
Lithospermum caroliniense, 223
liverleaf, 135
lobelia, 26
Lobelia siphilitica, 285
Lobelia species, 26
Lobelia spicata, 285
long-bracted spiderwort, 287
loosestrife, 237
loosestrife family (Lythraceae), 30
lopseed family (Phrymaceae), 32
Lotus corniculatus, 231
Lygodesmia juncea, 275
Lysimachia borealis, 237
Lysimachia species, 237
Lysimachia thyrsiflora, 237
Lysimachus, king of Thrace, 237
Lythrum salicaria, 313, 315

madder plant family, 33, 139
mad-dog skullcap, 311
Maianthemum canadense, 79
Maianthemum racemosum, 81
Maianthemum stellatum, 81
ma'kasîn (moccasin), 243
marsh, 13, 18
marsh hedgenettle, 171
marsh marigold, 251
marsh pea, 297
marsh skullcap, 311
Mather, Cotton, 123
Matricaria discoidea, 209
mayapple, 25
mayflower, 79
medicinal plants
- bluebeard lily, 227
- blue cohosh, 221
- Canada anemone, 127
- Canada mayflower, 79
- early figwort, 337
- great blue lobelia, 285
- large-flowered bellwort, 229
- long-bracted spiderwort, 287
- marsh skullcap, 311
- prairie ironweed, 279
- smooth rose, 185
- wild bergamot, 169
- wild cucumber, 87
- wood nettle, 269
- yellow lady's slipper, 243

melilot, 233
Melilotus albus, 233
Melilotus officinalis, 233
Mentha arvensis, 101
Mentha canadensis, 101
Mentha species, 29, 101
menthol, 101
mîcaogacan (odor of deer hooves), 69
Michigan lily, 193
mîdewidji'bîk (medicine lodge root), 127
"milk sickness", 65
milkweed, 23, 37, 151, 189
Mimulus ringens, 319
miner bees, 293
miner's lettuce family (Montiaceae), 30
mining bees, 201
mint family (Lamiaceae), 29, 309
Mirabilis jalapa, 173
Mirabilis nyctaginea, 173
Mitella diphylla, 143
Mitella nuda, 143
miterwort, 143
mitsuba, 51
moccasin flower, 175
Mohlenbrock, Robert, 10
monarch butterfly, 37
monarch caterpillars, 153, 189
Monarda fistulosa, 169
Monarda species, 29
monkeyflower, 319

Monotropa uniflora, 91
moraines, 13
morning glory family (Convolvulaceae), 27
mountain death camas, 107
mountain mint, 29, 103
mustard family (Brassicaceae/Cruciferae), 26, 71, 75, 225
Myosotis scorpioides, 25
myrsine family (Myrsinaceae), 30–31

naked miterwort, 143
narative, 41
narrow-leaved cattail, 339
narrow-leaved purple coneflower, 159
narrow-leaved wild leek, 47
natural communities, 17–19
Nelumbo lutea, 231
nettle family (Urticaceae), 35
New England aster, 277
New Guinea impatiens, 25
nodding bur-marigold, 203
nodding trillium, 145
northern bedstraw, 139
northern blue flag, 21, 305
Northern Lakes Ecoregion, 13–14, 16
northern red oak, 14
Northern Tallgrass Prairie National Wildlife Refuge, 16
Nuphar variegata, 239
Nymphaea odorata, 111, 239

ode'imîn (heart berry), 137
odîte'abûg (flat heart leaf), 239
Odysseus, 107
Oenothera biennis, 241
Ojibwe people, 47, 57, 69, 79, 87, 95, 127, 137, 169, 221, 227, 229, 239, 245, 269, 281, 283, 309, 311, 335
old man's nightcap, 83
o'makaki'wîdass "(frog's leggings), 335
onion, 149
onion family (Alliaceae), 23
orange flowers, 187–195
orchid, 177, 317
orchid family (Orchidaceae), 31
Oronoco Prairie Scientific and Natural Area, 16
Osmorhiza claytonii, 55
Osmorhiza longistylis, 55
Oxalis species, 247
Oxalis stricta, 247

Packera paupercula, 211
Packera plattensis, 211
paintbrush, 195
pale-spiked lobelia, 285
Panax quinquefolius, 61
paper birch, 13
Papilio polyxenes, 201
parsley, 23

parsnip, 23
peach, 33
peachleaf willow, 15
pea family (Fabaceae), 29, 293, 299
peatlands, 13
Pedicularis canadensis, 245
Pediomelum esculentum, 299
Pelargonium, 29
Pembina Trail Scientific and Natural Area, 16
Penstemon grandiflorus, 321
Persian violet, 28
Persicaria amphibia, 181
Petalostemum purpureum, 293
Philadelphia fleabane, 69
phlox family (Polemoniaceae), 32
phlox moth, 179
Phlox paniculata, 33
Phlox pilosa, 179
Phryma leptostachya, 119
pine, 141
pineapple-weed, 209
Pinguicula species, 29
Pinguicula vulgaris, 29
pink family (Caryophyllaceae), 26
pink lady's slippers, 175
pink to red flowers, 147–185
pipevine family (Aristolochiaceae), 24
pitcher plant family (Sarraceniaceae), 34
pitcher plant mosquito, 335
plantain family (Plantaginaceae), 32
plant family characteristics, 22–35
plant family classifications, 22
plant species, 21
plasterer bees, 293
Platanthera psycodes, 317
Platanthera species, 317
poison hemlock, 49
pollinators, 37, 157, 289
Polygonatum biflorum, 265
Polygonatum pubescens, 265
poppy family (Papaveraceae), 32
pothos, 261
prairie, 18
prairie blazing star, 163
Prairie Coteau Scientific and Natural Area, 16
prairie ironweed, 279
prairie larkspur, 133
prairie lily, 193
prairie milkweed, 153
prairie onion, 149
prairie phlox, 179
prairie ragwort, 211
prairie rose, 185
prairie turnip, 299
primrose, 241
primrose family (Primulaceae), 31

primrose moth, 241
private property, 16
Prunella vulgaris, 309
public properties, 16
Pulsatilla patens, 323
purple clematis, 131
purple giant hyssop, 307
purple loosestrife, 313, 315
purple pitcher plant, 9, 335
purple prairie clover, 293
purslane family (Portulacaceae), 30
pussytoes, 67
Pycnanthemum species, 29
Pycnanthemum virginianum, 103
Pyrola elliptica, 93
Pyrola species, 93

quaking aspen, 13, 15

radicchio, 273
radish, 75
ragweed, 215, 263
ragwort, 211
ramps, 47
Ranunculus hispidus, 253
Ranunculus species, 251
Ray, John, 21
ray flowers, 25
red baneberry, 125
red clover, 167
red maple, 13
red pine, 14
regal fritillary butterfly, 327
rein orchids, 317
Rheum rhabarbarum, 33
rhubarb, 33, 181
Rosa arkansana, 185
Rosa blanda, 185
rose, 33, 185
rose family (Rosaceae), 33
rose hips, 185
rough blazing star, 163
round-leaved sundew, 89
round-lobed hepatica, 135
Rudbeckia hirta, 213
rue anemone, 129
ruffed grouse, 61

Sagittaria latifolia, 45
sandalwood family (Santalaceae), 34
Sanguinaria canadensis, 117
Sarracenia purpurea, 9, 335
sarsaparilla, 61
savanna, 14, 19
saxifrage family (Saxifragaceae), 34
Schinia florida, 241
Schinia indiana, 179
scientific names, 21, 340–342
Scrophularia lanceolata, 337
Scutellaria galericulata, 311
Scutellaria lateriflora, 311
self-heal, 309
sharp-lobed hepatica, 135
Shauquethqueat, Joseph (Joe Pye), 161
shepherd's purse, 73

showy lady's slippers, 177
showy tick-trefoil, 295
Silene latifolia, 77
skeletonweed, 275
skunk cabbage, 261, 331
smartweed, 181
smooth oxeye, 207
smooth rose, 185
smooth Solomon's seal, 265
smooth yellow violet, 257
sneezeweed, 205
Socrates, 49
Solidago altissima, 215
Solomon, king of the Kingdom of Israel and Judah, 265
Solomon's plume, 81
Solomon's seal, 81
sour grass, 247
southern blue flag, 21, 305
sphagnum moss, 17
spiderwort, 287
spiderwort family (Commelinaceae), 27
spirea, 33
spotted Joe-Pye weed, 161
spreading dogbane, 57
spring, 19
spring beauty, 30
squash, 27
squirrel corn, 115
Stachys palustris, 171
Stachys pilosa, 171
starflower, 237
starfruit, 32
starry Solomon's plume, 81
St. Croix National Scenic Riverway, 16
St. Croix State Park, 16
strawberry, 33, 137
stream/river, 19
sugar maple, 14
sundew, 89
sundew family (Droseraceae), 27–28
sunflower, 24, 207
Superior National Forest, 13, 16
swamp, 13, 19
swamp aster, 277
swamp milkweed, 151
swamp thistle, 157
sweet cicely, 55
sweetflag, 199
sweetflag family (Acoraceae), 22–23
Symphyotrichum novae-angliae, 277
Symphyotrichum puniceum, 277
Symplocarpus foetidus, 331

tall goldenrod, 215
Tallgrass Prairie Ecoregion, 14–15, 16
tamarack, 13
Taraxacum officinale, 217
Tettegouche State Park, 16
Teucrium canadense, 171
Thalictrum thalictroides, 129

till plains, 13
timpsila, 299
titmice, 263
toothwort, 75
touch-me-not family (Balsaminaceae), 25, 269
Tradescantia bracteata, 287
Tragopogon dubius, 219
Trientalis borealis, 237
Trifolium pratense, 167
Trifolium repens, 99
Trifolium species, 167
trillium, 145
Trillium cernuum, 145
trillium family (Trilliaceae), 34–35
trout lily, 30
true forget-me-not, 25
true lily, 30
tufted loosestrife, 237, 313
Twin Lakes Scientific and Natural Area, 16
Typha ×glauca, 339
Typha angustifolia, 339
Typha latifolia, 339

Utricularia species, 29
Utricularia vulgaris, 235
Uvularia grandiflora, 229

veiny pea, 297
Verbascum thapsus, 255
verbena bee, 325
Verbena hastata, 325
verbena moth, 325
Vernonia fasciculata, 279
Veronicastrum virginicum, 123
vervain, 325
vervain family (Verbenaceae), 35
Viola communis, 327
Viola eriocarpa, 257
Viola pubescens, 257
Viola sororia, 327
violet, 257
violet family (Violaceae), 35
Virginia spring beauty, 109
Virginia waterleaf, 281
virgin's bower, 131
Voyageurs National Park, 16

waterfowl, 181
water hemlock, 49
water lily, 111, 239
water lily family (Nymphaeaceae), 31
watermelon, 27
water nymph, 111
water plantain family (Alismataceae), 23
watershield, 165
watershield family (Cabombaceae), 26
water smartweed, 181
wedding gown, 83
western ragweed, 263
white baneberry, 125
white camas, 107
white campion, 77

white cedar, 13
white clover, 99
white flowers, 43–146
white pine, 14
white snakeroot, 65
white sweet clover, 233
white-throated sparrow, 61
white trout lily, 105
white turtlehead, 121
wild bergamot, 29, 169
wild calla, 59
wild columbine, 183
wild cucumber, 87
wildflowers
  collecting, 37
  conservation, 10, 37
  ecoregions, 13–15
  guidebook tips, 39–41
  natural communities, 17–19
  sites, 15–16
  varieties, 9
wild four-o'clock, 173
wild garlic, 149
wild geranium, 303
wild ginger, 333
wild leek, 47
wild licorice, 97
wildlife, 181, 263
wild lily-of-the-valley, 79
wild mint, 29, 101
wild pea, 297
wild sarsaparilla, 61
wild strawberry, 137
wintercress, 225
Winthrop, John, 123
wood anemone, 129
wood betony, 245
woodland, 14, 19
woodland strawberry, 137
wood lily, 193
wood nettle, 269
wood sorrel, 247
wood sorrel family (Oxalidaceae), 32
woolly blue violet, 327
Wyeomyia smithii, 335

xawiska (flowers white), 91

yarrow, 63
yellow birch, 13
yellow flowers, 197–257
yellow lady's slipper, 243
yellow pond lily, 239
yellow rocket, 225
yellow sweet clover, 233
yellow toadflax, 249
yellow touch-me-not, 191
yellow trout lily, 105
yellow wood sorrel, 247

Zantedeschia species, 59
Zingiber officinale, 333
Zizia aptera, 201
Zizia aurea, 201
Zizia species, 201

Michael Homoya was the Heritage Program Botanist and Plant Ecologist for the Indiana Department of Natural Resources Division of Nature Preserves for 37 years before retiring in 2019. He is a fellow and former president of the Indiana Academy of Science and a board member and former president of the Indiana Native Plant Society. His previous books include *Orchids of Indiana*; *Wildflowers and Ferns of Indiana Forests*; *Wake Up, Woods* (a children's book); and *Wildflowers of the Midwest* (coauthored with Scott Namestnik). He was awarded the Distinguished Career Public Service Award from the Conservation Law Center, the Distinguished Scholar of the Year from the Indiana Academy of Science, and the Barbara J. Restle Lifetime Conservation Award from Sycamore Land Trust.